INDUSTRIAL
ENERGY SYSTEMS:
ANALYSIS, OPTIMIZATION
AND CONTROL

Richard E. Putman

Library of Congress Cataloging-in-Publication Data

Putman, Richard E., 1925-
    Industrial energy systems : analysis, optimization, and
control / Richard E. Putman.
    p. cm.
ISBN 0-7918-0208-6
1. Industries—Energy conservation.
    I. Title.
TJ163.3P88 2004
621.042–dc22   2003062834

To Agnes
Wife and Mother to Caroline and William
Best Friend and Constant Companion
for nearly 40 years and a
Fellow-Student in the University of Life

Acknowledgement

ASME Press would like to acknowledge the generous assistance of Richard Putman's colleagues and friends, Frederick C. Huff and George Saxon, Jr., in completing this important publication after the author's untimely passing. Without their kind efforts, this book may never have been published.

Tribute to the Author

Richard E. J. Putman, the author of this book, was known as Dick to his friends. Unfortunately, Dick passed away before this book was published. He died on Easter Sunday, April 20, 2003, at his residence in Charlotte, N.C., after a long and courageous battle with cancer. Dick was born in London, England on June 18, 1925. He received his degree in Mechanical Engineering from Paddington Engineering College, London, U.K., in 1945. Dick became a Chartered Mechanical Engineer in 1952. He was also a member of ASME and both the IEEE and ISA. Dick was elected Fellow of ASME in 2002.

Dick met his wife Agnes in Budapest, Hungary in 1961. They were married and shortly after Dick helped Agnes and her family flee from the former Communist Hungary. He went behind the Iron Curtain for her, persevering through the red-tape and danger, then secreting her out of Hungary, through Sweden, the UK, Australia, then Cleveland and finally to Pittsburgh. For the first two years that Dick and Agnes knew each other they had to communicate in Latin since Agnes did not know English and Dick did not speak Hungarian.

Between 1948 and 1964, Dick was engaged in the boiler control industry in the U.K. and Canada. From 1965 until 1990, he was an employee of Westinghouse where he became responsible for equipment performance analysis, advanced computer control technologies, and plant optimization strategies. He retired from Westinghouse in 1990 and became the Technical Director of Conco Consulting Corporation.

While at Westinghouse Dick became a pioneer in the field of Industrial Energy Management systems. During this time period he was awarded 38 U.S. patents to go along with his

6 British and one Canadian patent. He also authored more than 80 published technical papers and articles. Continuing his career with Conco, Dick became an expert in the areas of heat transfer and condenser performance.

Dick was always a gentleman and a technocrat of the highest order. While he worked at Westinghouse if there was ever a subject that the division needed an expert on Dick was always selected. He would give a technical explanation (deep enough to have lost you early-on) and would frequently close with a cheerful statement of, "..it's really quite simple". He never showed anger or frustration with his fellow employees or customers. He was so bright and articulate that he was always called on to talk to customers at every opportunity to help sell our products and services. He was reliable, always there, and was a truly positive influence on the people that worked with him. In the 1970's Dick was chosen to accompany former Westinghouse CEO Doug Danforth on a visit to China right after President Nixon opened trade between the US and China.

Dick excelled as a mentor and teacher. He always led by example and never asked any of his subordinates to do anything he wouldn't do himself. He willingly shared his knowledge and boundless enthusiasm with his co-workers. He freely offered his friendship and he really cared for the people with which he worked. There are many people today whose careers were greatly enhanced and whose lives were positively impacted by knowing and working with Dick.

After his formal retirement Dick continued to work as a consultant and an author. In 2000 the Electric Power Research Institute published "Condenser Inleakage Guideline," with Dick as a principal author, and in 2001 ASME Press published his book entitled "Steam Surface Condensers: Basic Principles, Performance Monitoring, and Maintenance".

The Latin expression "Carpe Diem" summarizes the way Dick lived his life. He seized every moment - he made the most out of every opportunity and day that he lived.

TABLE OF CONTENTS

Introduction   **xvii**

**Chapter 1   The Impact of Plant Economics on
             the Design of Industrial Energy Systems**   **1**

**Chapter 2   Managing Energy Resources from
             within the Corporate Information
             Technology System**   **7**

2.1   Introduction   7

2.2   Historical Perspective   8

2.3   Open System Architecture—Local
      Area Network (LAN)   13

2.4   Open System Architecture—Local
      and Wide Area Networks (LAN/WAN)   17

2.5   Connectivity Building Blocks for an
      Open System Architecture   20

    2.5.1   WAVE Server
                (Web Access View Enabler)   21

    2.5.2   ODBC Server for
                Open Data Base Connectivity   22

    2.5.3   OPC-OLE for Process
                Control (OPC)—Server   24

    2.5.4   NetDDE Server
                (Network Dynamic Data Exchange)   26

2.5.5   eDB Historian   28

2.6   Summary   29

Chapter 3   **Fundamental Characteristics of
            Energy Conversion Equipment**   **31**

3.1   Introduction   31

3.2   Fossil-fuel Fired Boilers   31

    3.2.1   The Calculation of Combustion
        Efficiency   34

    3.2.2   Estimating the Flow of a Fuel when
        its Flow is not Directly Measured   37

    3.2.3   Incremental Cost   39

3.3   Steam Pressure Reducing Valves   40

    3.3.1   Adiabatic Expansion   40

    3.3.2   Isothermal Expansion   43

3.4   Steam Desuperheaters   43

    3.4.1   Desuperheated Steam Temperature
        Set Points   46

    3.4.2   Steam Turbogenerators   49

    3.4.3   Simple Turbogenerator   49

    3.4.4   Back Pressure or Topping
        Turbogenerators   50

    3.4.5   Turbogenerators Controlled
        from Surplus Steam   53

    3.4.6   Extraction/Condensing Steam
        Turbogenerators   53

3.5   Steam Surface Condensers   59

    3.5.1   Steam Surface Condenser Operations   61

3.5.2 Steam Surface Condenser Maintenance   62

3.5.3 Steam Surface Condenser Tube
      Cleaning   62

3.5.4 Fouling Deposit Characteristics   65

3.5.5 Deposit Sampling   65

3.6 Cooling Towers   66

3.6.1 Cooling Tower Heat Transfer Theory   69

3.6.2 Cooling Tower Characteristic Curve   72

3.6.3 Predicting Cooling Tower Performance   74

3.6.4 Air Property Algorithms   75

Chapter 4   Optimization Strategies   **77**

4.1 Linear Programming Applications   80

4.1.1 Commentary   89

4.2 The Simplex Self-Directing Evolutionary
Operation (SSDEVOP) Optimizing
Technique   90

4.2.1 The Simplex Principle   92

4.2.2 SSDEVOP Experimental Design   93

4.2.3 SSDEVOP Optimization of a System
      with Two Turbogenerators   95

4.3 Steepest Ascent Method   98

4.4 Equal Incremental Cost Method   100

4.5 Optimal Trajectories   104

4.6 The Kalman Filter as a Process
Modeling Tool   106

4.7 Conclusions   109

Chapter 5   Applications and Case Studies   **111**

5.1   Linear Programming Applications   111

    5.1.1   Plant with Only One Steam
         Turbogenerator   111

    5.1.2   Paper Mill with Two Steam
         Turbogenerators   114

         5.1.2.1   An Off-Line Study   122

    5.1.3   Plant with One Condensing and
         One Backpressure Turbogenerator   123

         5.1.3.1   Costs   126

         5.1.3.2   Comments   126

    5.1.4   Optimal Utilization of Waste Heat
         Steam—Phosphate Processing Plant   128

         5.1.4.1   Variables   131

         5.1.4.2   Costs   132

         5.1.4.3   Optimization   132

         5.1.4.4   Comments   133

5.2   SSDEVOP Applications   135

    5.2.1   SSDEVOP Solution of Linear
         Problem   135

    5.2.2   SSDEVOP Solution of Non-Linear
         (Boiler) Problem   138

    5.2.3   Optimization of VARS within a
         Small Refinery Distribution System   142

5.3   Equal Incremental Cost Applications   152

    5.3.1   Range of Boilers   153

    5.3.2   Hydroelectric Turbogenerators   154

    5.3.3   In-plant Dispatching of Steam
         Turbogenerators   156

5.3.4   Optimal Load Trajectories   159

5.4   Combined Cycle Optimization   163

5.4.1.1   Data Flow   167

5.4.2   Equipment Models   170

5.4.2.1   Gas Turbogenerator   170

5.4.2.2   Heat Recovery Steam
          Generator   174

5.4.2.3   Steam Turbogenerator   177

5.4.3   Linear Programming Matrix   178

5.4.4   Results   182

5.4.4.1   On-Line Application   186

5.4.4.2   Accuracy of the Method   187

5.4.5   Optimization Results   188

5.5   Condenser/Cooling Tower Subsystem
      Optimization   190

5.5.1.1   Condenser Model   191

5.5.1.2   Cooling Tower Model   191

5.5.1.3   Commentary   192

**Chapter 6   Controlling the Steam Turbogenerator
             for Watts, VARS, Volts and Frequency   193**

6.1   Introduction   193

6.2   Turboalternators   196

6.3   Starting Up a Generator   200

6.4   Synchronizing Process   201

6.5   Single Steam Turbogenerator with No Tie-Line   202

6.6   Single Generator in Parallel with Tie-Line   207

6.7   Several Machines Operating in Parallel with
      No Tie-Line   209

6.8   Several Machines Operating in Parallel
      with Tie-Line Connected   210

6.9   Conclusion   210

Chapter 7   The Importance of Process Heat
**   Exchangers in Industrial Energy**
**   Systems   213**

7.1   Shell-and Tube Heat Exchangers   215

      7.1.1   Fundamental Thermodynamic
              Relationships—Shell-and-Tube
              Heat Exchangers   217

              7.1.1.1   Log Mean Temperature
                        Difference (LMTD)   218

              7.1.1.2   LMTD Correction Factor F
                        for a Two-Pass
                        Heat Exchanger   219

              7.1.1.3   Design Heat Transfer
                        Coefficient–U_{des} 220

              7.1.1.4   Tube Wall Resistance–R_w 222

              7.1.1.5   Fouling Resistance
                        Allowances–R_{fi} and R_{fo} 223

              7.1.1.6   Tube Side Film Heat
                        Transfer Coefficient–h_i 223

              7.1.1.7   Shell Side Film Heat
                        Transfer Coefficient–h_o 224

7.2   Heat Exchangers in the Alumina
      Extraction Process   225

7.2.1 The Fundamental Chemistry
of the Bayer Process   225

7.2.2 Bayer Process with
Red Mud Processing   226

7.2.3 Bayer Process
with Red Mud Rejection   230

7.3 Modern Heat Exchanger Maintenance
Procedures   231

7.3.1 Hydrodrilling Heat Exchanger Tubes   231

7.3.2 Hydrodrilling and High Pressure Water 233

7.3.3 Heat Exchangers with U-Tubes   233

7.3.4 Process Heat Exchangers   233

**Chapter 8   Demand Side Management
and Electrical Load Shedding   235**

8.1 Demand Control and Electrical
Load Shedding—Fixed Window   236

8.1.1.1 Predicting Demand Error
at the End of a Period   239

8.1.1.2 Types of Electrical Load   240

8.1.1.3 Principal Features of a Demand
Control System Strategy   243

8.1.1.4 Demand Control Stimulator   246

8.2 Demand Control and Electrical
Load Shedding—Sliding Window   246

8.3 Load Shedding in Response to
Plant Contingencies   248

8.4 Load Shed Speed and the Control System   252

Chapter 9	**Future Trends**	**255**
9.1	Real Time Electric Power Pricing	255
	9.1.1 Markets for Real-time Pricing	257
9.2	Neural Networks	258
Appendix A	**Mathematical Procedures**	**265**
A.1	Multiple Linear Regression Analysis	265
A.2	Numerical Analysis	266
	A.2.1 Fibonacci Search	267
	A.2.2 Regula Falsi Search	268
	A.2.3 Newton-Raphson Numerical Search Technique	268
A.3	The Newton-Raphson Technique Used in Equipment Performance Models	272
A.4	Linear Programs	274
	A.4.1 Discontinuous Linear Relationships	277
A.5	Using MicroSoft EXCEL^{TM} as "Solver" Feature for Linear Programming Problems	280
	A.5.1 Upper Constraints	281
	A.5.2 Equalities	284
	A.5.3 Lower Constraints	284
	A.5.4 Creating the EXCEL^{TM} Spreadsheet	284
	A.5.5 Completing Problem Initializing Using "Solver"	287
	A.5.6 Testing the Program	288
References		**289**
Index		**295**

Introduction

During the 1970s and 1980s, a body of knowledge was developed and applied to the on-line optimization and control of the energy systems to be found in industrial plants. The motivation for this activity was two-fold: (a) the cost of energy was relatively high so that by reducing the amount of energy consumed it was beneficial to the bottom line of the business and (b) the OPEC crisis of 1973 developed a general awareness that the conservation of energy was a laudable social goal and further helped to justify the installation of these systems. Today, the possible impact of energy emissions on global warming adds an additional stimulus to energy conservation.

Unfortunately, during the 1990s, the cost of energy fell to the point where management obtained a better return-on-investment on process productivity improvements so that there was a diminished interest in installing systems for the conservation of energy. As a result, there was a loss of familiarity in the U.S. with this technology and its application, on the part of both users and vendors. However, as we enter the Third Millennium, there are new pressures to conserve energy. The global warming debate has focused on the need to reduce carbon dioxide emissions while political instability in the Middle East has questioned the dependence on oil and its derivatives. As a result, those who manage energy in industrial plants in both developed and developing countries are being forced to re-evaluate the need for energy conservation.

This book is intended to capture the body of knowledge that was developed by a group of engineers within what was then the Process Control Division of the Westinghouse Electric Corporation and is now part of Emerson Electric. The techniques described have been applied successfully in many industries throughout the United States as well as in several industrialized countries in the Middle and Far East. The techniques for analysis can be especially useful in identifying possible avenues in equipment selection or flowsheet adjustments that would tend to reduce plant energy consumption without affecting the productive capacity of the plant. The term "optimization" describes a process in which a cost function is minimized through the wise selection of an operating strategy from among several. Thus, for optimization to be meaningful, there must be several valid ways of operating the equipment involved and its interconnections, each with a different cost. The tools described here will allow an analyst to evaluate the possibilities and identify the configuration that is most cost-effective. Other material in this book will help engineers who are responsible for the management of energy systems within their plants to apply techniques that have been used successfully elsewhere.

The writing of this book was greatly encouraged by George E. Saxon, Senior Chairman of Conco Systems, Inc. of Verona, PA. The support of Edward G. Saxon, President of Conco Systems Inc. is also greatly appreciated. The author would also like to express his indebtedness to those who have contributed to the text, especially Frederick C. Huff of Emerson Electric and George E. Saxon, Jr., Vice-President of Conco Systems. The original work was a team effort in which a group of engineers from a variety of disciplines enthusiastically worked together to build and apply this body of knowledge. This team has since become dispersed, but the author would like to take this opportunity to thank them all for their contributions.

Finally, no man is an island and the author offers heartfelt thanks his wife, Agnes, for her generous cooperation and for the time we might have otherwise spent together during the writing of this book.

Chapter 1
The Impact of Plant Economics on the Design of Industrial Energy Systems

Experience has shown that, at any point in time, the policy adopted by a plant for managing its industrial energy system varies with the relative price of energy, the energy resources that have been deregulated, the international political and economic climate and the energy policy currently being implemented by the national government. Thus the tools available for this management function must be capable of rapid adaptation to the set of energy resource conditions that currently apply. The rewards for a diligent pursuit of optimum operating strategies will also vary with these external factors. The energy resources can take many forms: electricity, steam, gas, fuel oil, byproduct fuel, etc. and may either be purchased or generated within the plant. Thus, tools for the analysis, optimization and control of the generation, consumption and distribution of multiple sources of energy are among those needed to perform this complex management function, taking account also of the interaction between the management of energy and its impact on the associated manufacturing process(es).

The task has grown more complicated over the years. Up until the 1970s, much of the activity focused on minimizing the fuel consumption of boilers and furnaces and maintaining adequate air/fuel ratios to prevent the formation of smog, the first having a direct economic impact on plant operating costs. It is interesting to note that, in 1916, the Hagan Corporation of Pittsburgh, PA, and the Bailey Meter Company in Cleveland, OH, developed the earliest combustion control systems and it was no coincidence that both cities were major centers of the U.S. steel industry, with a high concentration of furnaces belching smoke into the atmosphere. The maintenance of proper air/fuel ratios was also stimulated by two pollution incidents that occurred around 1953. One occured in Donora near Pittsburgh, PA and the other in the City of London. Both were the result of temperature inversions and the resulting smog killed hundreds of people. One response was the installation of combustion control equipment on all furnaces and boilers that did not already have them. Combined with other measures, such lethal smogs have not been experienced in either location since.

The economic and social benefits of combustion control technology were widely recognized after WWII and the technology was generally adopted by the utility industry, not only in the U.S. but overseas as well. Even at the end of WWII combustion control equipment specifically designed to function on mobile platforms could be found in the boiler rooms of most ocean going tankers, as well as in commercial and naval vessels.

Another energy management activity popular during the 1970s was the on/off control of electrical loads to take advantage of the rules embodied in utility power supply contracts that reflected some of the principles of power demand-side management. In addition to energy charges that depended on the time of day (off-peak or on-peak), they also included penalties for exceeding a demand limit, defined as the amount of power consumed during a 15- or 30-minute period converted to MW. The set of sheddable loads was identified and the switching priorities and operating constraints defined. Systems were designed to automatically switch the loads on and off so that not only was the demand limit not exceeded but violation of the constraints was also avoided. The early systems received a pulse from the utility at the beginning of each demand period but some companies found ways to game the system by drawing large amounts of power for short periods at the beginning of a particular period without violating the demand limit at the end of the period. Utilities then switched to what became known as the "sliding window" approach and installed a new type of recording demand meter, the demand limit being redefined as the amount of energy consumed during any 15- or 30-minute period experienced at any time during the day.

In 1973, the price of oil rose dramatically in response to events in the Middle East but Presidents Nixon and Ford were able to control the domestic price of oil and gas. At the time, some interest developed in applying energy optimizing strategies to industrial plants but the return on investment (ROI) was not yet significant enough. Lovins (2001) notes that the fall of the Shah of Persia in 1979 again hiked oil prices but, by this time, President Carter had deregulated oil and gas prices, which rose together with the price of coal. This time, the shock was sufficient to stimulate a nation-wide, seven-year drive for greater energy efficiency. The price of industrial process control computer systems had also fallen and economical ways were found to apply a number of different on-line optimization strategies, successfully reducing the cost to industrial plants of the fuels and electric power that they had to purchase.

One of the policies introduced under President Carter was to increase the efficiency of American built cars by seven miles per gallon over six years. As a result, oil imports from the unstable Persian Gulf region fell. Between 1977 and 1985, while the US GDP rose 27%, oil imports fell by 42% and caused the OPEC countries to lose one-eighth of their market. In 1985–86 world oil prices collapsed but this had the unfortunate effect of reducing the ROI from financing energy conservation projects and, in the U.S., new capital investments shifted to projects designed to improve manufacturing productivity. However, energy costs in developing countries remained high and for the rest of the decade they retained a continuing interest in installing new energy conservation projects.

The terrorist attack on the World Trade Center in New York in September 2001 created a new and troubling situation. Earlier that year the national energy policy proposed by the Bush administration encouraged the development of renewable energy sources, but was more focused on increasing domestic oil supplies combined with the increased use of coal and natural gas as the fuels for power generation. The development of safer nuclear power also formed an important part of the new policy. Unfortunately, OPEC was intent on maintaining a high world price for oil by cutting back production as needed. Furthermore, the economies of some of these countries depended on the industrialized countries maintaining, or even increasing, their oil consumption. As a result, the complexities of the Middle East situation precluded attempts by the U.S. to reduce oil consumption and even attempts to raise automobile efficiency were dropped.

The situation was further complicated by the effects of deregulation of the electric power industry that also occurred at this time. In some parts of the country, the spot price of electric power sometimes fluctuated wildly and it became necessary for large users to enter into long-term contracts. These price fluctuations were attributed to questionable practices on the part of energy traders that also sometimes affected the price of natural gas. Other uncertainties include the future effect of reorganizing the power transmission industry and the introduction in some areas of the "real-time pricing" of power.

A further uncertainty is associated with the concept of global warming. There is no doubt that atmospheric carbon dioxide does trap heat within the biosphere but whether an observed increase in global temperatures is due to man-made emissions or to more natural phenomena (e.g. the sun cycle) is in dispute. Some even question whether the temperature increases are really occurring. As a result, the United States has withdrawn from the Kyoto Treaty. However, some of the signatories are prepared to adopt an emissions trading scheme that will allow those companies that have exceeded their commitments to receive monies from those who wish to purchase emissions credits to compensate for missing their targets. How this will effect a particular plant is not yet clear but this cost/revenue possibility has to be taken into account.

Clearly, as we enter the Third Millennium, the economics on which future energy planning is to be based present a very confused picture. For this reason, it is essential that the tools and techniques used to manage energy resources allow current energy prices to be reflected in the strategy and solution. Most of the basic strategies were developed over the period 1973–1990 and have been documented in numerous papers presented to learned societies as well as in patents granted by the U.S. Patent Office during that period. The purpose of this book is to capture the philosophy and method on which this work was based in the expectation that it will prove to be invaluable to a new generation of engineers as they wrestle with these problems.

Chapter 2

Managing Energy Resources from within the Corporate Information Technology System

2.1 Introduction

The deregulation of the electric utility and natural gas industries and, in some localities, the advent of "real-time pricing" for electric power, have introduced a new dynamic into the decision-making processes that determine how these utilities are purchased and managed within an industrial plant. The process becomes even more complicated when an enterprise wishes to manage these utilities at the corporate level for a number of their manufacturing plants. While the optimum energy distributions within each plant can be determined independently, the results are very dependent on the instantaneous prices of the various purchased utilities, each of which may be determined by decisions made at the corporate level and, in some circumstances, can also vary rapidly with time. This has had an impact on the architecture of the local control and data acquisition systems that are located within each plant and has required the addition of interfaces between the plants and the centralized Information Technology systems used at the corporate level.

As a result, the design of modern distributed Control Systems (DCS) is driven by the hardware and software concepts embedded in Information Technology (Yeager

(1997)). Control platforms have evolved from the closed systems that used highly proprietary, vendor-specific components, each with their own customized operating systems; towards integrated and open systems that use a variety of standard off-the-shelf hardware and software products. Until recently, standard, commercially available computer technologies simply could not offer the guaranteed response times, multi-tasking capabilities, redundancy, and industrial reliability necessary for mission-critical control applications. That situation has changed rapidly. Many off-the-shelf, commercially available hardware platforms offer performance capabilities that equal or even surpass the reliability and functionality offered by conventional proprietary distributed control systems and programmable logic controllers (PLC's). In addition, the off-the-shelf operating systems can now provide multi-tasking and real-time features that make them suitable for control applications.

Also, application software developed for business and engineering markets, such as relational database management systems (RDBMS), spreadsheets, and computer-aided design (CAD) programs are finding a home in the control room. This migration of commercial desktop- computing products into the previously closed world of control technology led to the next logical step; applying the emerging Internet/intranet technologies to unify control and business computing systems into one enterprise-wide platform, spanning from the process sensor to the boardroom.

2.2 Historical Perspective

The evolution of digital technology within the power and process industries has been marked by several sequences of first embracing a current system architecture, and exploiting it to its limit of capability and reliability: then changing the architecture to avoid the roadbump, providing a new direction to the evolutionary path. From the beginning, application engineers insisted that the systems perform in real time and be highly reliable.

It is over 40 years ago that the digital control of industrial processes first became a reality. Among the computers available at that time were a number of models manufactured by both IBM and the Control Data Corp, complete with rudimentary input/output devices connecting the processors to the signals from analog electronic instruments mounted in the plant. A pioneering system was manufactured by the Control Data system and installed at the Little Gypsy plant in Louisiana; it is now in the Smithsonian Museum. Of other systems manufactured by Control Data, one was installed at the Sterlington, LA plant. One of these systems became famous when a technician used a high voltage instrument to check external circuit continuity and blew every transistor in the machine. As a result, the designers of input/output devices introduced the concept of optical isolation between the plant and the computer so as to avoid similar disasters in the future.

In spite of such setbacks, the process industries took a keen interest in the development of the new digital technology since it was generally recognized that analog technology had exhausted its possibilities. In 1959, the interest evolved into a forward-thinking movement when the Instrument Society of America sponsored that year's Power Conference in Kansas City. Attended by power plant engineers from throughout the U.S., it was dominated by the visionary, Bill Sommers of Ebasco. The papers presented at the conference outlined the possibilities presented by the new technology in improving the quality of control of power plant boilers and turbines, improved man/machine interfaces, extensive data acquisition and alarm features and much more.

In the early sixties, a number of equipment manufacturers became concerned how the new digital technology would be used to protect their own equipment, as well as that the new technology would perform in real time and also exhibit high systems reliability. As a result, a group of capable companies such as General Electric and Westinghouse began to develop their own digital control systems. Major analog control equipment manufacturers, including Honeywell and Foxboro, also began to follow this approach. These manufacturers took advantage of the advances in solid-state technology and the ability to purchase both chips and microprocessors, developing a number of simple real-time processors manufactured and programmed to their own specifications. In addition to interfacing to the plant instrument and control signals, these processors included multi-level interruptible task schedulers. One of the pioneers of this genre was the P-50 industrial process computer introduced by Westinghouse Electric Corp. in 1964 and based on transistor technology, with a 14-bit memory and programmed in assembly language. Others soon followed. Unfortunately, all of these early systems had limited memory capacity due to cost and were difficult to program.

Higher level languages were developed to overcome the programming difficulty. Most were based on FORTRAN 77 but others were interpretable languages in which the control functions were defined and linked to the process I/O through compilers. The control algorithms were also redesigned in accordance with sampled-data theory and both absolute and velocity algorithms were introduced. The latter had the advantage that, since they only transmitted changes, a failure of the control system would cause the changes to be suspended, so preventing the plant from suddenly being moved to an undesirable state before the operator could intervene.

As the cost of memory became reduced, the amount of memory grew. However, the early microprocessors were comparatively slow and it became difficult to maintain the real-time nature of the systems. One response was to design systems with multiple processors and shared core memories. These systems, introduced in the late seventies, provided additional levels of on-line redundancy, which improved their reliability, but they were very complex.

A parallel development at that time was distributed control, in which a number of independent control devices were provided, each with its own microprocessor and set of input/ output devices. To create a system with this built-in redundancy, these devices communicated with each other via high speed data links or even a common data highway driven by a host computer. Developed in the early eighties, this equipment was the forerunner of today's distributed control systems that now use high-speed fiber optic data highways.

A typical distributed control system (DCS) of the late eighties is illustrated in Figure 2.1. The unifying feature was the pair of redundant coaxial data highways that were either in the form of rings or else were open-ended cables that ran through the plant. Proprietary data link protocols were used to handle the data transmission task. The input/output devices of various types, both local and remote, were mounted in the plant and connected to process control units, each of which had a pair of redundant microprocessors that scanned the data and broadcasted it over the data highways in real time. The process control units also performed the closed loop process control functions programmed into them.

Within an air-conditioned control room were located a number of operator consoles, an engineers work station, a calculator for performance and other computations, a historian to archive the plant data over a period of time, as well as one or more printers. These constituted the windows through which the operators supervised the plant, and monitored both control system and plant performance.

All of these systems were dedicated to the measurement and control of plant variables and only on rare occasions were they

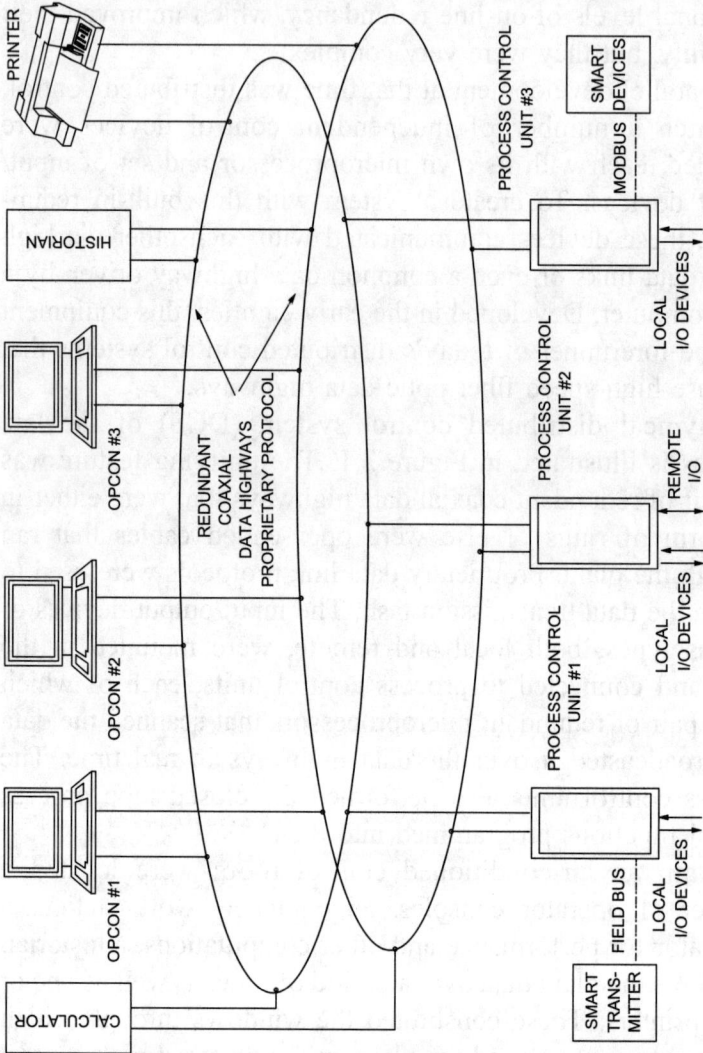

Figure 2.1 Typical distributed control system (DCS) architecture

connected by a data link to the computer systems used to manage other parts of the business. However, by the end of the eighties, relational data base software became generally available and offered the possibility of incorporating some management functions within the process control system. Among these functions were data analysis, model generation, spreadsheet presentations and maintenance planning. Unfortunately, systems supplied by different vendors were often incompatible with one another and the task of integrating a variety of subsystems into one integrated system was both formidable and costly. A variety of data transmission protocols were also involved, many of them proprietary, so increasing the complexity of the integration task.

2.3   Open System Architecture—Local Area Network (LAN)

The nineties saw a movement towards systems with an open architecture, free from the incompatibilities commonly experienced up to that time. The new systems, catering to the needs of both operations and management, were built around database software such as Oil Systems™ or Oracle™, used commercially available hardware including personal computers, and were able to communicate with the distributed control systems still needed for the control of the process itself.

Brownlee (2001) reviews some of the open standards for communications in use at that time, including serial communications protocols RS-232 and RS-485, which were comparatively slow. Before open architecture could be achieved, Yeager (1997) refers to the number of new data transmission standards that had to be created and adopted by the computer and control equipment vendors. Among these were:

FDDI   Fiber Distributed Data Interface
CDDI   Copper Distributed Data Interface
TCP/IP   Transmission Control Protocol/Internet Protocol
ATM   Asynchronous Transfer Mode
Ethernet IEEE 802.3 Standard
Fast Ethernet

Until recently, FDDI has been the most widely employed technique. Based on ANSI standard X3T12, it has been successfully applied in numerous real-time, mission-critical business and avionics systems. It operates at 100 Mbps (megabits per second) and can transmit over distances of up to 200 kilometers.

A typical plant system based upon open system architecture is indicated in Figure 2.2, which shows the configuration of a system located within a plant but unable to communicate with a corporate IT system located elsewhere. In this case, the central feature of the system consists of redundant circular fiber optic FDDI data links using TCP/IP protocol, these constituting a local area network (LAN). The input/output system is completely separate and has its own redundant fiber optic communications links based on Ethernet protocol. The data obtained from the plant devices are passed to the FDDI data links by means of a pair of routers, which make the conversion from Ethernet to FDDI protocol while also providing a firewall between the two major subsystems. The development platform, operator station and historian are all connected to the FDDI highway, together with the process controllers, all this equipment being located in the plant control room. Furthermore, a geographical positioning system (GPS) server, receiving date and time signals from the GPS satellite, can also be connected to the FDDI highway so as to provide

DEVELOPMENT
PLATFORM

OPERATOR
STATION

HISTORIAN

REDUNDANT
CIRCULAR FIBER OPTIC
FDDI LINKS WITH
TCP/IP PROTOCOL

GPS
SERVER

ROUTER #1

ROUTER #2

FIREWALL

SWITCH — SWITCH

REDUNDANT
ETHERNET
FIBER-OPTIC
COMMUN.
LINKS

PROCESS
CONTROLLER
#1
(WITHOUT I/O)

PROCESS
CONTROLLER
#2
(WITHOUT I/O)

SWITCH — SWITCH

LOCAL I/O

LOCAL I/O

LOCAL I/O

SATELLITE
DISH

I/O
POINTS

I/O
POINTS

I/O
POINTS

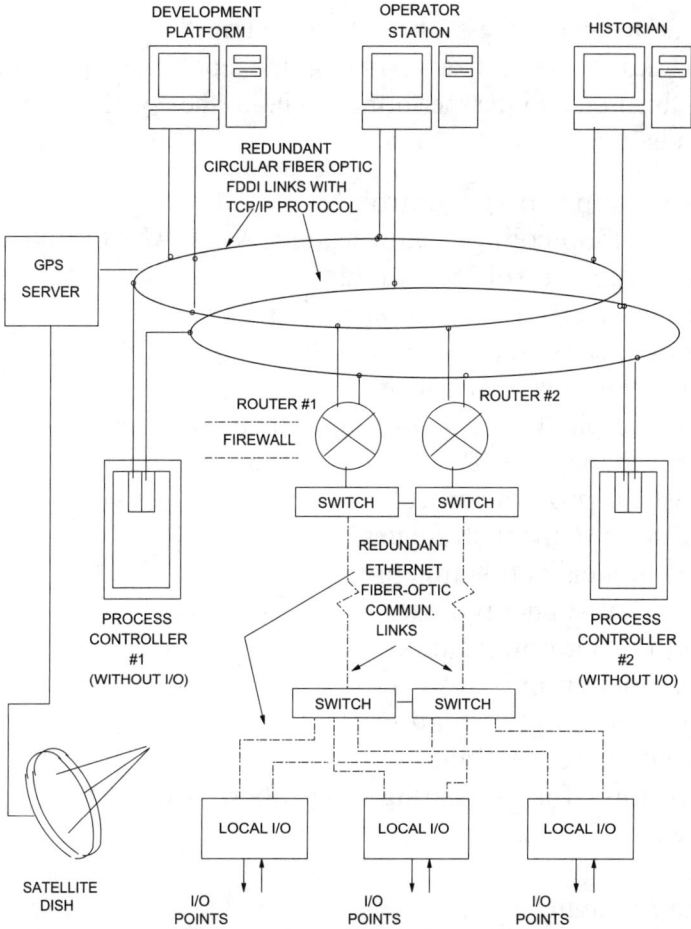

Figure 2.2   Typical open architecture control system

the system with an independent and continuous update of
current time. This has avoided the need for manual synchronization of any internal system clocks.

The early nineties also saw a considerable expansion in the set
of control and management functions to be implemented within
an industrial plant. A plant may contain several distinct processes each of which requires real-time data-related functions

unique to that process, the data including the various energy resources being utilized. Among the control and monitoring functions to be performed within the system may be included:

Closed loop process control
Control loop configuration using CAD/CAM techniques
Automatic control loop tuning
Automatic electrical load shedding
Process performance calculations
Equipment health monitoring
Dynamic plant model with instrument response
Alarm management
Statistical process control
Advanced control strategies
Plant models and status prediction
Controllable energy losses
Sootblowing optimization
Emissions optimization
Cogeneration optimization
System maneuverability
Algorithms for calculating the properties of
    steam, air, gas, etc.
Data logging
Data archiving

Some of these application functions involve advanced process control technology, including the programming of neural networks for adaptive process model generation (Zink 1999) used in some forms of process optimization (e.g. emissions optimization). Others involve enhanced data processing features (e.g. statistical process control and alarm management), general optimization technology or merely access to utility functions such as the sets of algorithms needed to compute the properties of air, steam, gas, etc., etc. A wide range of application skills and disciplines is clearly involved and the procurement of a set of these functions that are internally compatible with one another is a task in itself.

2.4   Open System Architecture—Local and Wide Area Networks (LAN/WAN)

Consider a paper manufacturing company that has three plants, geographically separated, and that it wishes to supervise them from a central location, integrating the plant data into its Information Technology system. Figure 2.3 not only shows how a wide area network (WAN) can be used by a centralized corporate IT system, located elsewhere, to communicate with the in-plant configuration of plant A as shown in Figure 2.2; but it is also able to use the WAN to communicate with plants B and C.

Over recent years, the capabilities of Fast Ethernet Fiber Optic LAN's have been developed to the point where they equal or exceed those of the FDDI LAN's that were the original choice for use with open architecture systems. As a result, there has been a tendency for the Fast Ethernet LAN's to replace FDDI technology both on account of cost and on their increasingly widespread availability among vendors. Otherwise, the configuration of in-plant systems is very similar to that shown in Figure 2.2. The main difference would be, for instance, the addition of a WEB gateway that enables the data contained in the in-plant system to be accessed remotely by the corporate IT system over the Internet. Alternatively, Preheim et al. (1998) describe an alternative form of Wide Area Network based upon microwave communications, using transceivers and microwave towers to communicate with the IT center.

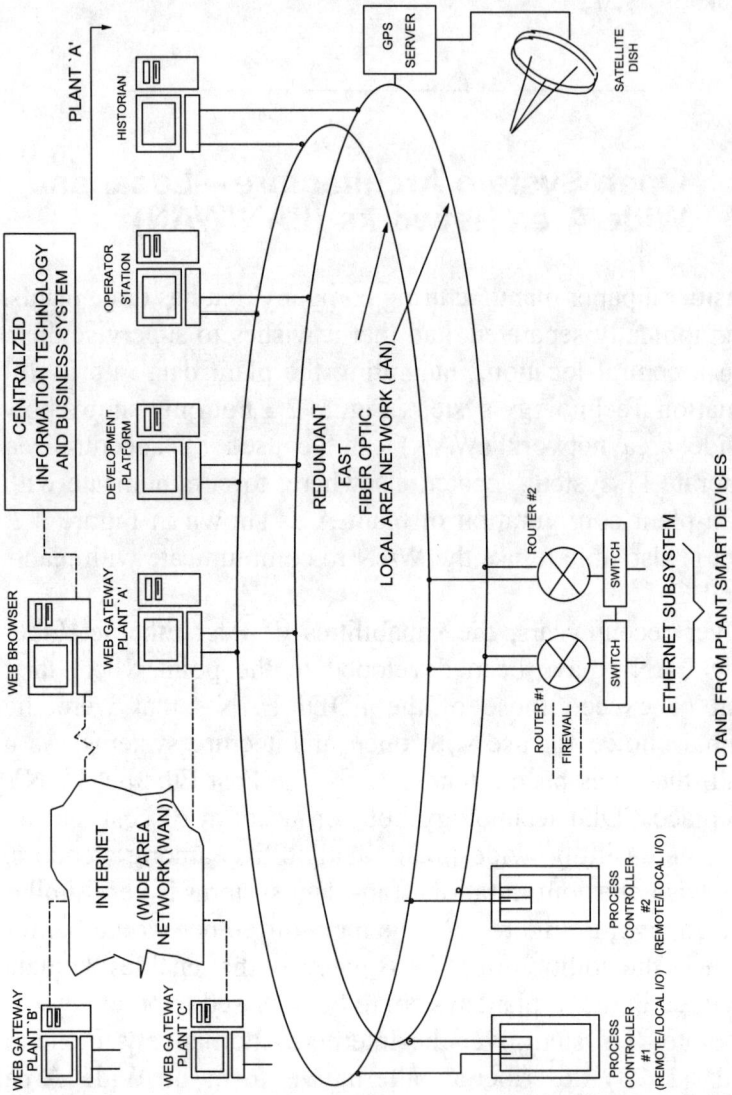

Figure 2.3   Open architecture system with WAN and LAN

Whatever the system employed, the data base management software used in the corporate IT system has to be made compatible with the software used in the in-plant systems. The adoption of the Standard Query Language (SQL) by almost all data base system vendors is one way to ensure such compatibility. There is also a rule that data can only be accessed (or queried) by the data base manager so that the system is generally organized not to allow any plant data to be changed. This avoids any external and unexpected interference in the control of a process without the knowledge of the operators. However, some enterprises may perform the energy optimization task at the IT center and then download a group of optimized control system setpoints to each plant for implementation. Open system architecture has allowed the required degree of flexibility to be introduced in such situations without distorting the standardized software that has been developed for these systems.

Some of the application functions that may be resident in the corporate IT system include:

Emissions monitoring
Statutory emissions reporting
Regulatory compliance
Preventive maintenance analysis and maintenance scheduling
Life extension analysis
Monitoring the real time pricing of power and fuels; organizing and implementing the response
Development of power trading applications
Centralize the optimization strategies for remote plants
Human factors engineering
Diagnostics
Contract calculations
Spreadsheet analysis
Data security
Virus protection

The lists are continuously expanding as new functions become desirable or necessary.

2.5   Connectivity Building Blocks for an Open System Architecture

The in-plant systems based on open systems architecture rely upon the set of standard data transmission protocols to perform their functions. However the arrangement of the various groups of stored data may not be compatible with one another and a set of connectivity building blocks had to be developed in order to facilitate the use of the data by the IT network These building blocks have become available in the form of servers that use personal computer hardware that is available from many vendors as off-the-shelf items. Their functional characteristics are contained in the customized software that is written for them, although the input and output communications systems use the standard data transmission protocols referred to in Section 2.2 above. Often the server is a device located in the plant and connected to the local data highway. However, the server software can be loaded into an existing device with which it is closely related. For instance, the ODBC software may be loaded into the historian shown in Figure 2.3.

Thus a number of other software standards have had to be developed to facilitate inter-connectivity, among them being:

WAVE   – Web Access View Enabler
ODBC   – Open Data Base Connectivity
OPC   – OLE (Object Linking and Embedding) for
             Process Control
NetDDE   – Network Dynamic Data Exchange
eDB   – Historian

Note that while the terminology used in describing the following important servers sometimes draws upon the terms used by both Westinghouse Process Control™ and Oracle™, the functionality described for these important functions is common to many vendors of open architecture systems.

2.5.1 WAVE Server (Web Access View Enabler)

The WAVE server provides graphic process data to any desktop computer with an Internet browser that is directly connected to the company WAN. The WAVE server is the Web Gateway shown in Figure 2.3 and, on one side, is connected to the dual redundant LAN located in the plant while the other side has a telephone connection to the Internet. It consists of only a Java-capable web browser such as Netscape Navigator or Microsoft Internet Explorer for a PC Workstation and no software resides permanently in this desktop computer. The WAVE server also includes a Cisco Firewall to ensure process security between the control system and an enterprise level LAN/WAN connection. Note that, because all access to plant information is read-only, control functions are disabled so that no proprietary hardware or software is required for the WAVE client workstation.

The hardware used for the WAVE Server may be a dedicated Windows NT PC or Sun Ultrastation connected to either the company LAN/WAN or to the IT data base Network. When the user wants to view plant process information, the web browser in the PC workstation located in the IT center requests the information from the WAVE server located in the plant. This accesses live DCS data over the LAN and transfers the data in the form of Java™ applets from the plant network directly to the desktop. By using the Internet to deliver plant process data to any location, WAVE servers reduce support costs as well as response times. On-call support professionals can trouble-shoot plant control systems from any location, without being required to travel to the plant.

The graphic display abilities embodied in WAVE in fact create a virtual "read-only" IT operator workstation. Although not all of the process diagram features are enabled in the WAVE Server, the available graphics are identical to the standard graphics that are embedded in the distributed control system, so providing a familiar point of reference for plant personnel. The diagram conversion utility available with WAVE is used to transform standard DCS trend displays, charts, text, point information, and plant process data so that the plant data can be viewed remotely through a Web browser. All live data on the diagram is updated periodically – typically every two seconds.

2.5.2   ODBC Server for Open Data Base Connectivity

Prior to the development of ODBC, there was no recognized standard for requesting plant *process* data and no way to connect the user to the data source. Data access, if requested across different systems, required customized software. Open Data Base Connectivity (ODBC) software has thus become the standard for connecting applications to both local and remote databases and provides standardized access to critical historic data. A server based on ODBC allows historical plant process data to be integrated directly with any type of ODBC-enabled application, examples of which include off-the-shelf word processors, spreadsheets and database application development tools. The ODBC works with the Structured Query Language (SQL), which has emerged as the standard language for accessing and filtering data. With this combination of standard software, it is possible for applications to access the historical databases in different systems without requiring the use of a proprietary data link. This standardized approach reduces development costs and ensures a reliable, easy to use interface to access critical historic plant performance data.

The ODBC server is connected on one side to the plant LAN and on the other side to the WAN to which the central IT system is connected. It imports historic process data directly from the historian to a given desktop computer, eliminating the effort and error of manual data entry. Performance data from the distributed control system (DCS) historian can be combined with accounting or other information, allowing plant managers to compare expected plant performance. The DCS data can also be combined with laboratory analysis data, Maintenance Management Systems (MMS) and Lab Information Systems; all can access plant data without a custom-designed interface.

Note that the ODBC should be used for applications that run on demand and is the preferred retrieval method when the data is to be stored in a relational database.

Some forms of ODBC server use a UNIX-based processor that contains ATI's *Open*RDA® and Database Access Manager® (DAM) components embedded in the operating system. This allows users to build new ODBC applications using standard database development tools. Because many development languages such as Visual Basic, C, C++ and Basic already have ODBC SQL capabilities, ODBC provides an easy way for plants to develop new customized ODBC client applications. There are multiple ODBC client applications already available for 16-bit Windows 3.1 (requires a TCP/IP protocol stack), and for 32-bit Windows NT or Windows '95. A sample of ODBC-enabled third-party applications include:

Microsoft Visual Basic
Microsoft Access
Microsoft Excel
Microsoft Query
Powersoft Powerbuilder
Oracle Forms

Oracle Reports
Gupta SQL Windows
Borland Delphi
C, C++
Microsoft MFC
Intersolv Q+E
Crystal Reports
Cognos Impromptu

2.5.3   OPC - OLE for Process Control (OPC) – Server

The OPC Server collects data from a field device via the plant
network or highway and delivers it to the OPC client
software. It is based on the use of the real-time data interchange protocol known as Object Linking and Embedding
(OLE) for Process Control, an open standard that provides a
consistent method of accessing data from plant devices. OLE
(Object Linking and Embedding) for Process Control was
designed by an open non-profit foundation to meet the general
needs of industry.

The OPC Server bridges the gap between third-party Win-
dows-based applications and process control hardware. OPC
servers can easily and automatically access dynamic process
information, so integrating the data from third-party applications with the data base manager used by the process control
system. These predefined standards allow any OPC client
software to communicate with any OPC server, regardless of
the type of device. However, it is not possible to access
historical data or events and alarm information through the
OPC server.

The OPC standard also enables users to easily integrate
plant data with plant chemical analysis applications, regardless
of the type and source of the data. The OPC does not require a
custom interface or driver in order for an external application

to access a plant's process control device. OPC server software can be developed once and can then be reused by any business, SCADA, HMI or custom software package. The OPC specification provides for future developments in technology and functionality, allowing OPC components to meet the emerging needs of industry.

OPC-compliant software applications, such as advanced control products, typically reside on the plant's business LAN/WAN, or on a user's local drive. Widely used PC tools such as spreadsheets and database management systems are already OPC-compliant, providing a standard way to import data from external data sources. Plants can also implement OPC client applications for advanced control using, for example, the SmartProcess™ Global Performance Adviser or Enterprise Process Historian.

The OPC point browser organizes the thousands of process points in a tree-like structure, allowing easy access to all plant point information by both the client application and the end user. The OPC Server accesses DCS data by creating groups and by assigning specific process points to each group. The OPC client application determines how frequently the application will check the DCS for new point information. A single message from the server to the client application contains the values for any points in the group that have changed. Because the points are grouped, and because the user determines how frequently the data is updated, the speed and performance of the OPC server is typically faster than NetDDE systems (see below).

Because client applications can use differing values such as floating point vs. integer values, the OPC Server automatically converts the data type if the client and server conflict. Further, because many development languages such as Visual Basic for Applications (VBA), C++ and Basic already have OPC interface routines, the development of new customized OPC client applications for a plant being greatly facilitated.

Some of the features of an OPC Server include:

- Standard OPC Client applications such as the GPA from SmartProcess™ can read current plant information for plant performance analysis
- Provides the ability to use a third-party OPC compliant software to send new point information on the DCS highway
- Provides continuous and automatic updating of information without specific user request
- Transfers data to client applications on an exception basis to maximize throughput
- Time stamps and assigns a quality to data (GOOD, FAIR, POOR, BAD and TIME-OUT) to inform the operator regarding the reliability of the data
- Interfaces between different types of DCS systems, allowing an operator to see all the appropriate set of process information
- Supports data access requests only. It is not possible to access historical data or events and alarm information through the OPC server.
- Maintains a log of system errors and major operations in a user-concied location. This log can be used to diagnose problems and analyze plant performance
- Writes information pertaining to any major program faults to a Microsoft Windows NT event log.

2.5.4 NetDDE Server (Network Dynamic Data Exchange)

An extension to Microsoft's DDE (Dynamic Data Exchange), NetDDE is a protocol to transfer data between programs across the LAN/WAN, without a custom interface or driver. It is a real-time data interchange protocol that is based on an open standard and another consistent method of accessing data from plant devices.

NetDDE exchanges information between a NetDDE server and NetDDE compliant software applications called "clients". A NetDDE server collects data from a field device via the plant network or highway and delivers it to the NetDDE-enabled software application that can reside on the plant's business LAN/WAN or on a users local drive. A NetDDE server eliminates the need for a proprietary data link protocol. It also provides an easy way for plants to develop new customized NetDDE client applications without the use of special software or toolkits.

The NetDDE Server is typically an existing device within the distributed control system that has been concied to function as a NetDDE Server. This device collects data from a field device via the plant network or highway, and then delivers it to the NetDDE-enabled software residing in the network or desktop PC. PC's or client workstations are connected to the LAN/WAN and allow users to access these client applications. Typical architecture for use with a NetDDE Server may be an Ovation Network.

NetDDE-enabled software applications, such as Microsoft Excel, Microsoft Access or other plant analysis tools, typically reside on the plant's business LAN/WAN, or on a users local drive. Widely used PC tools such as spreadsheets and database management systems are already NetDDE-enabled, providing a standard way to import data from external sources without requiring the use of a proprietary data link protocol.

NetDDE should be used for applications that require a steady stream of data.

Among the features available with the NetDDE Server are:

- Reads current plant information for plant performance analysis using external NetDDE-enabled applications such as Microsoft Excel, Microsoft Access, etc.

- Writes new point information on the DCS highway using third-party NetDDE-enabled software
- Transfers 100 data items each second, providing continuous update of information without specific user request
- Maximizes throughput by transferring data to client application on an exception basis
- Provides interfaces between different types of DCS systems, enabling an operator to see all process information regardless of its source
- Maintains a log of all system errors and major operations in a user-configured location. This log can be used to diagnose problems and analyze plant performance
- Writes any major program faults to the Microsoft Windows NT Event Log

2.5.5   eDB Historian

This eDB historian has been implemented as an Oracle DBMS application optimized for real-time data collection. Although it may be employed as a plant historian, the eDB supports distributed sources and users of the historical data located throughout a geographically dispersed organization, so as to become an enterprise process information system.

The eBD Historian is tightly integrated with the distributed control system for the collection of point values and status data, alarms, operator actions, and sequence-of-event messages. The eDB also supports the periodic acquisition of process values from third-party systems. This historical information is available from, for example, Oracle tables via SQL/ODBC queries to provide data analysis tools and user applications with either an aggregate or high-resolution view of the process activity. In this way, users are able to drill-down to uncover the precise data behind a plant event or calculation result.

The eDB includes a graphical trend function as well as a fully featured reporting system. Reports may be scheduled for periodic execution or to be initiated by a user or a particular plant event. The output may be in printed form or distributed in one of a variety of electronic formats, including Adobe Acrobat PDF or HTML.

Desktop applications (e.g. spreadsheets and data bases), business systems and custom programs have easy secure access to the eDB historical data tables via the SQL interface built into Oracle. Powerful and complex data requests may be formed to extract process data from other plant variables based upon specified plant criteria.

2.6   Summary

The purpose of this chapter has been to explore how the data from several plants may be integrated into the corporate Information Technology system, the need for an open architecture in order that this should be accomplished and some of the standard hardware and software that is available to achieve this. However, it is clear that the design of the system and its associated network(s) must be assigned to specialists who are familiar with the range of standard hardware and software that is available and how it may be used. The flexibility that has been incorporated in the software developed for use with an open architecture system also means that it is possible to customize the application without distorting the standards.

Chapter 3

Fundamental Characteristics of Energy Conversion Equipment

3.1 Introduction

Before a system can be analyzed, the behavior of the components of that system must first be understood. In this chapter, a number of the most important components of an Industrial Energy System are studied and their basic heat and mass equations are provided, together with their energy conversion equations, when appropriate.

3.2 Fossil-fuel Fired Boilers

Industrial power houses that practice cogeneration usually generate the live steam in boilers fired with one or more fossil fuels. A study of boiler behavior requires that the analyst has available at least the combustion efficiency vs. load relationship for each of the fuels used, as well as the unit cost of that fuel in $/MBTU. ASME Power Test Code PTC.4-1998 contains a detailed method for calculating combustion efficiency for a variety of modern power plant boilers but, with multi-fuel firing, does not distinguish between the combustion efficiency of each fuel being fired. For industrial boilers, the simpler test procedure contained in PTC.4.1-1974 has been found quite adequate for estimating and comparing their combustion efficiency; and, when multiple fuels are fired but the flows of some (e.g. bark) are not directly measured, the insights provided by Buna (1956) allow the mass flows of these fuels to be calculated individually.

Most of the heat released by the combustion process is used to raise the enthalpy of the feedwater supplied to the boiler to that of the steam leaving the boiler. However, some of the heat is lost to the process, the losses being identified, in BTU/lb fuel, as:

LG = Heat leaving the boiler contained in the dry gases that are the products of combustion

L_{mf} = Heat contained in the fuel moisture as it leaves the boiler

L_H = Heat contained in the moisture created by the combustion of the hydrogen in the fuel

L_{ma} = Heat leaving the boiler contained in the moisture supplied with the combustion air

L_r = Heat loss due to radiation through the boiler walls (See Figure 3.8 of ASME (1974))

$$\text{Total losses} = L_{TOT} = L_G + L_{mf} + L_H + L_{ma} + L_r \quad (3.1)$$

If the higher heating value of the fuel (HHV) is known in BTU/lb fuel, and it should be known at the time the fuel is purchased, then combustion efficiency η_c as a percentage is defined as:

$$\eta_c = \frac{HHV - L_{TOT}}{HHV} * 100 \quad (3.2)$$

Figure 3.1 shows typical relationships between combustion efficiency and load for various fuels and it will be noted that, for a given fuel, efficiency first rises as the load is increased, up to a maximum known as the maximum continuous rating (MCR) for the boiler. However, should the load be increased still further, the efficiency will tend to fall off slightly.

MAXIMUM CONTINUOUS RATING
(MCR)

COAL

BLAST FURNACE GAS

NATURAL GAS

COKE OVEN GAS

COMBUSTION EFFICIENCY - %

BOILER LOAD - LB/H

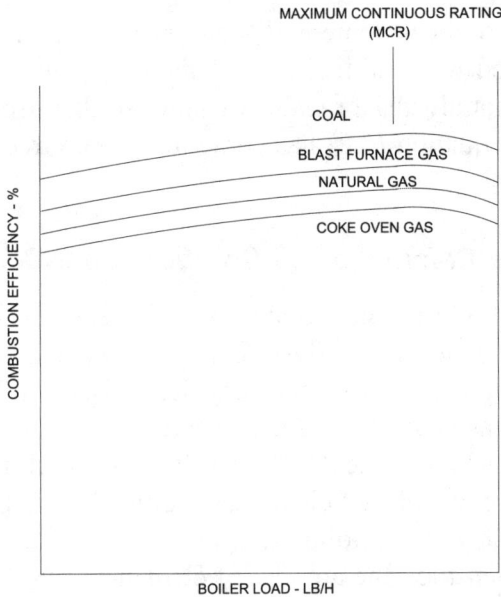

**Figure 3.1   Typical boiler efficiency vs. load curves multi-fuel
fired boilers**

Experience shows that it is not difficult to fit the data used to
generate these efficiency vs. load curves to a second or third
order polynomial in terms of load (W_s), using standard
regression techniques, thus:

$$\eta_c = a_0 + a_1 W_s + a_2 W_s^2 + a_3 W_s^3 \qquad (3.3)$$

Note that combustion efficiency is not the same as output-input
efficiency in which the heat acquired by the steam is divided
by the total amount of heat supplied with the fuel. Some of the
heat in the latter is included in the boiler blowdown or in the
atomizing steam (if any) but these losses are not included in
equation (3.2). However, these costs have a comparatively
small effect on the optimum distribution of load and fuel mix
among a range of boilers and, if desired, can be accommodated
by making an appropriate adjustment to the fuel price. In any

case, due to measurement difficulties, the total amount of heat supplied with the fuel used in the output-input efficiency method is not always a reliable quantity so that use of the loss method of equation (3.2) has come to be preferred.

3.2.1 The Calculation of Combustion Efficiency

The calculation of boiler combustion efficiency is complicated by the fact that fuel analysis data is presented in terms of percent weight per lb of fuel or, alternatively, BTU/lb fuel; while flue gas analysis data is presented as percent by volume of the various constituents. Bearing this in mind, the calculation of combustion efficiency as outlined in ASME(1974) proceeds generally as follows:

For a given fuel, the amount of elements C, S, H, O and N will be given (lb/lb); as will the flue gas analysis in terms of volume percentage of CO_2, CO and O_2.

Use Column 3 of Table 3.1 to calculate the weight of nitrogen corresponding to the theoretical air required for the combustion of C, S and H in the fuel.

Use Column 8 of Table 3.2 to calculate the volume (V) of the flue gases when only the theoretical amount of air is being supplied.

The amount of excess air that corresponds to the O_2 concentration by volume (O_X) measured in the flue gases is now calculated. If V_{O2} is the volume of oxygen and V_{EA} the volume of excess air in the flue gases, and the concentration of oxygen in air is 21% by volume, then:

$$V_{EA} = V_{O2}/0.21$$

and

$$\frac{O_X}{100} = \frac{V_{O2}}{V + V_{EA}}$$

Table 3.1   Combustion properties of major fuel elements

Element	Atomic Weight	Combustion Product	1 Density of Gaseous Prod.	2 Lb Air/ Lb. Element	3 Lb N_2/ Lb. Element	4 Lb Product/ Lb. Element
C	12.01	CO_2	0.1170	11.4847	8.8202	12.4847
S	32.0666	SO_2	0.1730	4.30147	3.30352	5.30147
H_2	1.008	H_2O	0.0476	34.209	26.2725	35.209
O_2	16.000	—	0.0846	—	—	1.000
N_2	14.008	—	0.0744	—	—	1.000
Air	—	—	0.0765	—	—	—

Notes:

1. Density in lb/ft³ dry gas at 60°F and 30 in Hg
2. $W_{AC} = 2*O/(0.232*C)$ (0.232 is weight fraction of oxygen present in air)
   $W_{AS} = 2*O/(0.232*S)$
   $W_{AH} = O/(0.232*2*H)$
3. $N_2 = (1.0 - 0.232)*$(lb air/lb element in Column 2)
4. Prod = (lb air/lb element in Column 2) + 1.0

Table 3.2   Volumetric Properties of Major Fuel Elements

Notes	5	6	7	8
		Lb Gas/Lb Element		Cu.ft Gas/
Element	CO_2	SO_2	N_2	Lb Element
C	3.6644		8.8202	31.3196
S		1.9979	3.30352	11.54855
H_2			26.2725	—
N_2			1.0000	13.44086

Notes:

5. Subtract Column 3 from Column 4 of Table 3.1 for carbon
6. Subtract Column 3 from Column 4 of Table 3.1 for sulfur
7. Same as Column 3 of Table 3.1
8. Divide Column 5 by gas density in Column 1 of Table 3.1

   Divide Column 6 by gas density in Column 1 of Table 3.1

   Divide 1.0 by gas density in Column 1 of Table 3.1

from which

$$V_{O2} = (V * O_X)/(100 - O_X/0.21) \qquad (3.4)$$

From Column 1 of Table 3.1, density of oxygen is 0.0846 lb/ft^3, so that weight of excess air:

$$W_{EA} = (0.0846 * V * O_X)/$$

$$(0.232 * (100 - O_X/0.21)) \qquad (3.5)$$

The total mass of flue gas per lb. fuel may now be calculated using the constants in column 4 of Table 3.1, thus:

$$W_{TOT} = W_{EA} + 0.124847 * C + 0.0530147 * S$$

$$+ 0.262725 * H \tag{3.6}$$

Note that only the nitrogen associated with the combustion of H_2 is included in the flue gas flow, the moisture vapor being ignored.

The five heat losses identified earlier, namely L_G, L_{mf}, L_H, L_{ma} and L_r may now be calculated in BTU/lb of fuel, using the equations defined in ASME(1974), from which L_{TOT} can be computed. Boiler combustion efficiency may then be calculated using equation (3.2).

3.2.2 Estimating the Flow of a Fuel when its Flow is not Directly Measured

Having established the combustion efficiency of each fuel (η_{ci}) of n fuels being fired, the mean efficiency for the boiler η_{mean} is a weighted average, calculated from:

Let W_i = weight of fuel (i) being fired lb/h
   H_i = heat content of fuel (i) BTU/lb
   η_{ci} = combustion efficiency of fuel (i) being fired decimal fraction

Then

$$\eta_{mean} = \frac{\displaystyle\sum_{i-1}^{n} H_i W_i \eta_c i}{\displaystyle\sum_{i=1}^{n} H_i W_i} \tag{3.7}$$

However, when industrial boilers are fired with coal or bark, their fuel flow is difficult to measure. Buna (1956) outlined a calculation method that allows a reasonable estimate of the unmeasured fuel flow rates to be obtained even with multi-fuel firing. The method was developed in the days when the Orsat apparatus was the conventional method for flue gas analysis. Although the test gave the concentrations of CO_2, CO and O_2 in the flue gases, the CO_2 was the sum of both the SO_2 and CO_2, expressed as a percentage. Let

C_M = Carbon in fuel   Percent
S_M = Sulfur in fuel   Percent
H_M = Hydrogen in fuel   Percent
O_M = Oxygen in fuel   Percent
N_M = Nitrogen in fuel   Percent
a_M = Ash content of fuel   Percent
CA = Carbon in ash decimal fraction
H_{ACQ} = Heat acquired by steam BTU/h
CO_2 = Concentration of CO_2 in flue gases   Percent
CO = Concentration of CO in flue gases   Percent
O_2 = Concentration of oxygen in flue gases   Percent

Then from a gas balance:

$$Z = K(C_{bM} + 0.369S_M) + O_M - 0.7937H_M$$

$$-0.302N_M - 0.014S_M = 0.0 \qquad (3.8)$$

Where

$$C_{bM} = C_M - a_M CA$$

$$K = \frac{100 - 4.78CO_2 - 2.88CO - 4.76O_2}{1.42CO_2 + 1.41CO}$$

and

$$\sum_{i-1}^{n} W_i Z_i = 0.0$$

If only W_1 and W_2 are not measurable, then the following can be written:

$$W_1 Z_1 + W_2 Z_2 + \sum_{i=3}^{n} W_i Z_i = 0.0$$

from which

$$W_1 Z_1 + W_2 Z_2 = -\sum_{i=3}^{n} W_i Z_i \qquad (3.9)$$

and

$$W_1 W_1 \eta_1 + W_2 H_2 \eta_2 = H_{ACQ} - \sum_{i=3}^{i-n} W_i H_i \eta_i \qquad (3.10)$$

Clearly, solving simultaneous equations (3.9) and (3.10) allows the values of W_1 and W_2 to be calculated.

3.2.3 Incremental Cost

It was mentioned earlier that data reflecting combustion efficiency and load can be regressed as a second or third order polynomial as shown in equation (3.3). However, if the cost of fuel (i) is to be calculated from the boiler load, then:

$$Cost_i = W_{si} Hsi / \eta_i$$

If $1/\eta$ is regressed against load Ws, then with a second order fit,

$$Cost_i = W_{si} H_{si} (a_0 + a_1 W_{si} + a_2 W_{si}^2)$$
$$= H_{si} (a_0 W_{si} + a_1 W_{si}^2 + a_2 W_{si}^3)$$

and incremental cost

$$\Delta_{cost} = H_{si}(a_0 + 2a_1 W_{si} + 3a W_{si}^2) \qquad (3.11)$$

Clearly, the roots of equation (3.11) represent the two loads which have the same incremental cost Δ_{cost}, and indicates the rich opportunities offered by optimizing strategies based on the equal incremental cost principle. A similar result is obtained if the data for $1/\eta$ vs. load is regressed using a third order fit.

3.3   Steam Pressure Reducing Valves

A steam pressure reducing valve is provided with a source of live steam and is designed with a variable area orifice (e.g. needle valve) that allows the corresponding steam flow to pass through the valve but exit at a lower pressure. As shown in Figure 3.2, the expansion process takes place in two stages:

3.2.1 At constant entropy as the steam expands through the orifice from point (A) to point (B) and

3.2.2 Overall isothermal expansion, i.e. at constant heat, in that the enthalpy of the steam leaving the valve (point (C)) is essentially the same as that entering the valve (point (A)), subject to a frictional loss of about 10% (Lewitt (1953)).

3.3.1   Adiabatic Expansion

When steam expands through an orifice there is a maximum pressure drop which, if exceeded, causes the orifice to become choked. If the pressure at the inlet to the valve is P_{in}, then the

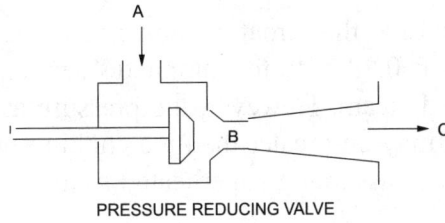

PRESSURE REDUCING VALVE

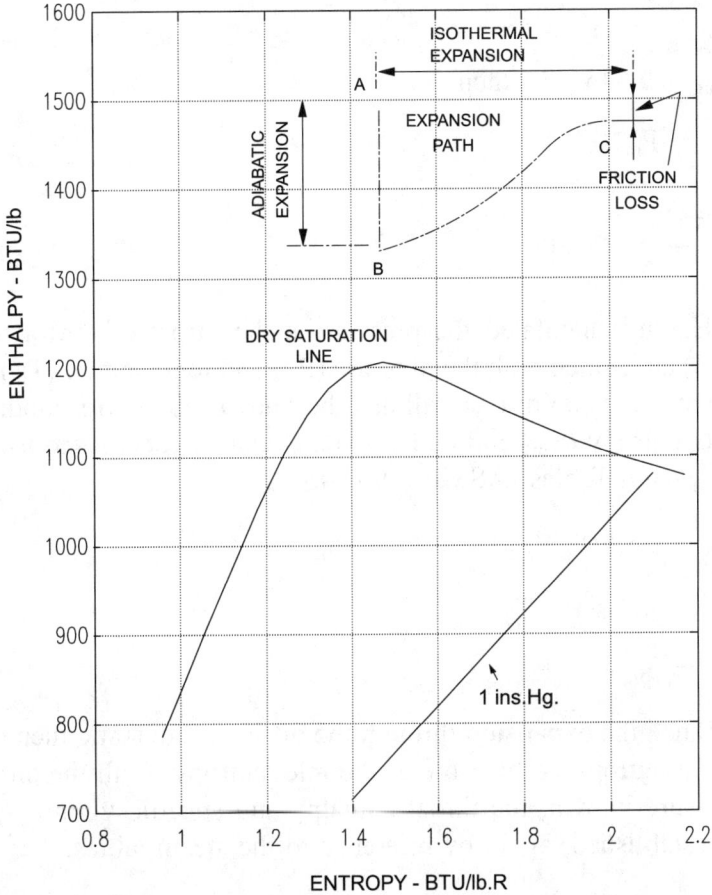

Figure 3.2   Steam expansion through a pressure reducing valve

choking pressure at the throat of the valve orifice is given by Lewitt (1953) as $0.58 * P_{in}$ for saturated steam and $0.545 * P_{in}$ for superheated steam. However, the pressure at the outlet of the valve P_{out} may be greater than the choking pressure at the throat P_{thr}. Thus, assuming superheated steam:

If

$$P_{out} > 0.545 * P_{in} \text{ then}$$

$$P_{thr} = P_{out}$$

Else
$$P_{thr} = 0.545 * P_{in} \tag{3.12}$$

Having calculated the pressure at the throat of the orifice (P_{thr}) and, since both the pressure (P_{in}) and temperature (T_{in}) of the steam at point (A) will also be known, the corresponding inlet entropy (S_{in}) and enthalpy (H_{in}) can be established using the Steam Tables (ASME (1967)):

$$H_{in} = stbfh(P_{in}, T_{in})$$

$$S_{in} = stbfhs(P_{in}, T_{in})$$

$$S_{thr} = S_{in}$$

Since the expansion through the orifice is adiabatic then the throat entropy is the same as the inlet entropy. With the throat pressure known, the throat enthalpy and specific volume can be established, again by reference to the steam tables.

$$T_{thr} = stbft(P_{thr}, S_{thr})$$

$$H_{thr} = stbfh(P_{thr}, T_{thr})$$

$$SPV_{thr} = stbfV(P_{thr}, T_{thr})$$

Lewitt shows that the throat velocity V fps can be calculated from the difference in enthalpy between the inlet and the throat, thus:

$$V = 224\sqrt{H_{in} - H_{thr}} \qquad (3.13)$$

If W_{thr} is the mass steam flow at the throat (lb/h), then the throat area in square inches can be calculated from:

$$\text{AREA} = 144 * (W_{in} * SPV_{thr})/(V * 3600) \qquad (3.14)$$

3.3.2   Isothermal Expansion

Line BC of Figure 3.2 indicates the path that the expansion follows after the steam leaves the orifice. The actual trajectory will depend on the shape of the discharge chamber, the design of the skirt of the valve plug and related matters. However, the essential feature is that the enthalpy of the steam at the discharge pressure is almost the same as that at the inlet to the valve. Thus, and again referring to the steam tables:

$$H_{out} = H_{in}$$

$$T_{out} = stbft(P_{out}, H_{out})$$

The importance of being able to calculate T_{out} is that it can now be determined whether any desuperheating may be required in order to reduce the temperature of the steam discharged from the valve down to that of the steam in the header which receives it.

3.4   Steam Desuperheaters

As suggested in the previous paragraph, steam desuperheaters may be needed to reduce the temperature of the steam discharged from a pressure reducing valve down to the conditions of the steam desired in the outlet header. Desuperheaters may also be installed prior to a process to protect the equipment, to improve the heat transfer properties of the steam or to assist its condensation.

Figure 3.3 shows the significant mass flow and enthalpy parameters involved and it is assumed that the desuperheating water mixes thoroughly with the cooled steam and is, thus, evaporated rapidly. It will also be noted from Figure 3.3 that desuperheating normally takes place at a sensibly constant pressure. If W_{in} and H_{in} are the mass flow rate and enthalpy of the steam to be cooled, W_{ds} and H_{ds} are the mass flow rate and enthalpy of the desuperheating water, and H_{out} is the desired enthalpy of the desuperheated steam, then:

$$(W_{in}H_{in}) + (W_{ds}H_{ds}) = (W_{in} + W_{ds})H_{out}$$

from which

$$R_{ds} = (W_{ds}/W_{in}) = (H_{in}-H_{out})/(H_{out}-H_{ds}) \qquad (3.15)$$

The ratio R_{ds} is usually termed the desuperheating water ratio. It will be shown that in the course of energy system analysis, the desuperheating water ratio can be used to calculate header steam balances without requiring that a separate variable be used to denote the desuperheating water flow rate itself. See commentary on Table 3.1 below.

Condensate drawn from a condenser is the ideal source of desuperheating water. It is free from the dissolved solids that tend to deposit out and obstruct the smaller orifices or spray nozzles through which the water must pass. If another source of water must be used, it is likely that regular cleaning will be required in order to remove the deposits.

Some desuperheaters are designed with nozzles that rely only on the kinetic energy of the water to create a finely

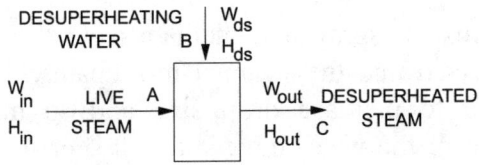

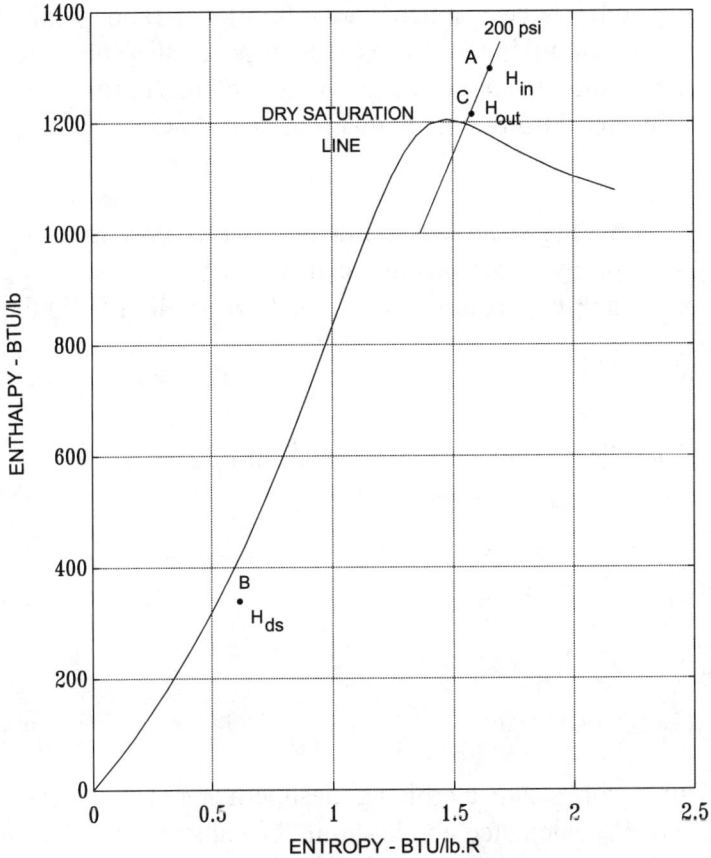

Figure 3.3   Steam Desuperheater

divided mist or spray that facilitates the evaporation of the water injected into the steam. Unfortunately, low water flow rates tend to increase drop size and so make the rapid absorption of the water more difficult. For this reason, some desuperheaters are provided with a source of higher pressure steam that is injected with the water and helps to vaporize it. Such nozzles have a much wider turndown ratio, while the amount of atomizing steam used is quite small. The effect of the atomizing steam on the properties of the desuperheating fluid injected into the main steam is as follows:

Let

k = atomizing steam/desuperheating water flow ratio
H_{as} = Enthalpy of atomizing steam BTU/lb
H_{df} = Enthalpy of combined desuperheating fluid BTU/lb

Then

$$(W_{ds} * H_{ds}) + kW_{ds} * H_{as} = (1+k)W_{ds} * H_{df}$$

Or

$$H_{df} = (H_{ds} + kH_{as})/(1+k) \qquad (3.16)$$

and

$$W_{df} = (1+k)W_{ds} \qquad (3.17)$$

Thus, with steam atomizing desuperheaters, the values of H_{df} and W_{df} calculated above should be substituted for H_{ds} and W_{ds} in equation (3.15)

3.4.1 Desuperheated Steam Temperature Set Points

In the process industries it was long thought that effective heat transfer in heat exchangers was possible only if the steam was supplied at close to saturated temperature. In this case, condensation would occur immediately the steam entered the heat exchanger and the response of control loops would be fast, with minimal time delays. There were also good economic reasons for this belief:

- Delivering steam close to saturation temperature would minimize the amount of live steam that has to be delivered to the process steam system, the difference being supplied by adding inexpensive desuperheating water, or condensate, to satisfy the total process demand.
- The amount of heat lost during transmission through the steam distribution system would also be minimized.

However, controlling the temperature of the process steam close to saturation is not easy, even with modern control equipment. Thus, when stability problems are experienced in trying to run at only a few degrees above saturation, engineers reluctantly tend to raise the set point of the temperature control loop, the reluctance based on a concern that increasing the degrees of superheat is detrimental to heat transfer performance. However, if unstable operation occurs when the temperature falls close to saturation, it can lead to a suspension of feedback, causing large amounts of condensate to collect in puddles on the bottom of distribution system piping. If the traps have difficulty in removing the large quantity of condensate, the steam piping can become damaged, sometimes from corrosion fatigue. There is a further concern that temperature cycling can reduce the life expectancy of the equipment.

Over the years, a body of research has emerged which suggests that the film heat transfer coefficient tends to rise with an increase in the degrees of superheat. This means that the temperature of the process steam can be allowed to rise without negatively affecting the heat transfer characteristics but with significant improvement in the accuracy of control, especially under changing load conditions.

Minkowycz and Sparrow (1966) show an 8% increase in heat transfer rate with 100 degrees of superheat at common process steam operating pressures. Meanwhile, equations provided by Hewitt, Shires and Bott (1994) show about a 5% improvement for 50 degrees of superheat. The film heat transfer coefficient can be calculated from the following:

Let h_{nu} = Nusselt film heat transfer coefficient (BTU/(sq.ft.h.Deg.F)

h = Adjusted Nusselt film coefficient (BTU/(sq.ft.h.Deg.F)

ξ = Correction factor

C_{pG} = Specific heat of vapor at constant pressure

T_G = Temperature of vapor phase Deg.F

T_{sat} = Saturation temperature corresponding to the pressure of the vapor Deg.F

h_{LG} = Latent heat of condensation from saturated vapor to liquid phase BTU/lb

$$\text{Film heat transfer coefficient} = h = h_{nu}(1 - \xi)^{0.25} \qquad (3.17)$$

$$\text{where} : \xi = \frac{C_{pG}(T_G - T_{sat})}{h_{LG}} \qquad (3.18)$$

O'Keefe (1986) has detailed the engineering considerations involved in the design of desuperheating systems for power and process plants. It is a complicated subject and the quality of control is affected by flow turndown in addition to the setting of the temperature controller. It should also be noted that raising the temperature of the desuperheated steam at one point can have negative consequences for those customers receiving steam after its pressure has been further reduced.

3.4.2 Steam Turbogenerators

The steam turbogenerators installed in industrial plants are available in a number of different configurations. Some are used just to generate power and have the characteristics of machines installed in power plants. Some simple machines are used to control the pressure in a process steam distribution header, the drop in steam enthalpy being used simultaneously to cogenerate power. Other more complicated machines are arranged to extract process steam from intermediate stage(s) of the turbine, the associated steam enthalpy drop again being used simultaneously to cogenerate power. Frequently the low pressure stage of the turbine on these extraction machines exhausts into a condenser, allowing the steam to be recovered in the form of condensate, together with its heat.

The cogeneration of process steam and power has certainly been standard practice in most major industries in the United States since WWII, among these being the steel, pulp and paper, chemical and petroleum industries. The inclusion of condensers has also provided greater flexibility to operations management by allowing them to choose from a wider range of steam and power distribution assignments, especially when several extracting/condensing machines are operating in parallel. For instance, tie-line capacity and energy charges often vary throughout the day: by appropriately managing these condensing machines, producing more condensing power when tie-line energy rates are high, the total cost of plant electrical energy can be minimized. Unfortunately, condensing machines are not quite as common in industrial plants located overseas so that these offer fewer opportunities for significant energy savings.

3.4.3 Simple Turbogenerator

A simple turbogenerator is defined as one receiving live steam from a boiler or range of boilers and generating power only. Some steam may be extracted from the turbine to preheat the boiler feed water before it enters the boiler but the characteristics of the turbogenerator may be defined in terms of the unit heat rate, a typical plot of which is shown in Figure 3.4.

There are three definitions of unit heat rate, the units of which are BTU/kwh:

1. Turbogenerator Net Unit Heat Rate - BTU's acquired by the steam per hour divided by the net kilowatts delivered to the bus bars
2. Turbogenerator Gross Unit Heat Rate - BTU's acquired by the steam per hour divided by the gross kilowatts generated, including the power driving the unit auxiliaries
3. Overall Unit Heat rate – (BTU's acquired by the steam per hour divided by boiler efficiency) divided by the net kilowatts delivered to the bus bars

Definition (1) is used when several turbogenerators are operating in parallel and receiving steam from one common boiler or a range of boilers. Definition (3) is used when two or more turbogenerators are operating in parallel and each has its own dedicated boiler.

A variation of this type of turbogenerator occurs when the exhaust steam supplied to the condenser is used to heat a source of hot water that is circulated locally to residences and industry. These are known as combined heat and power plants (CHP) and are very popular in Europe.

3.4.4 Back Pressure or Topping Turbogenerators

Back pressure (or topping) turbogenerators are machines which receive steam from a suitable source but the governor is designed to control the flow of steam to maintain the pressure at the exhaust from the turbine at a sensibly constant

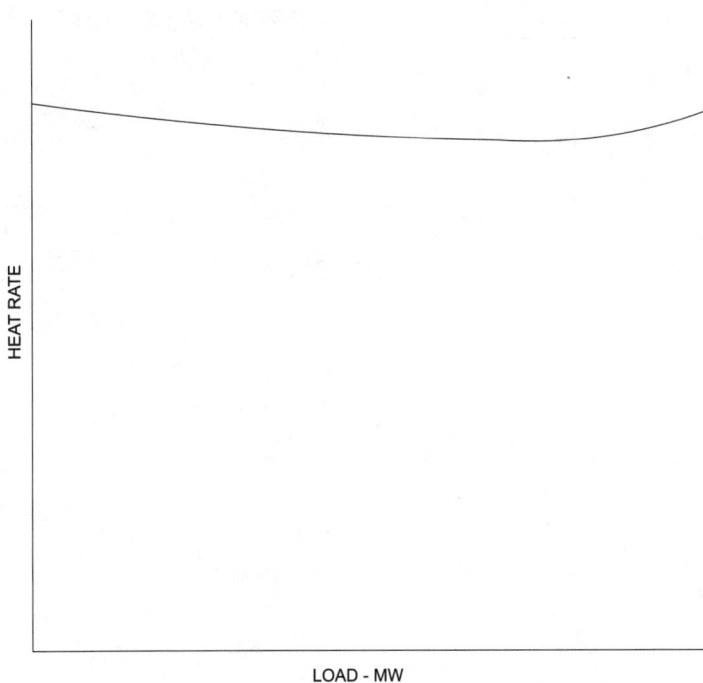

Figure 3.4   **Turbogenerator heat rate vs. load**

value (Figure 3.5). Such machines act essentially as a steam pressure reducing valve but have the advantage that the flow of steam that passes through them also generates electrical power. This cogeneration characteristic reduces plant operating costs but the amount of power generated can not be determined independently since it is entirely dependent on the back pressure deviation or its equivalent in terms of steam flow.

The relationship between power generation and steam flow is normally represented as a straight line generally as shown in the lower part of Figure 3.5. This is often termed the Willans Line and is a fundamental property common to all

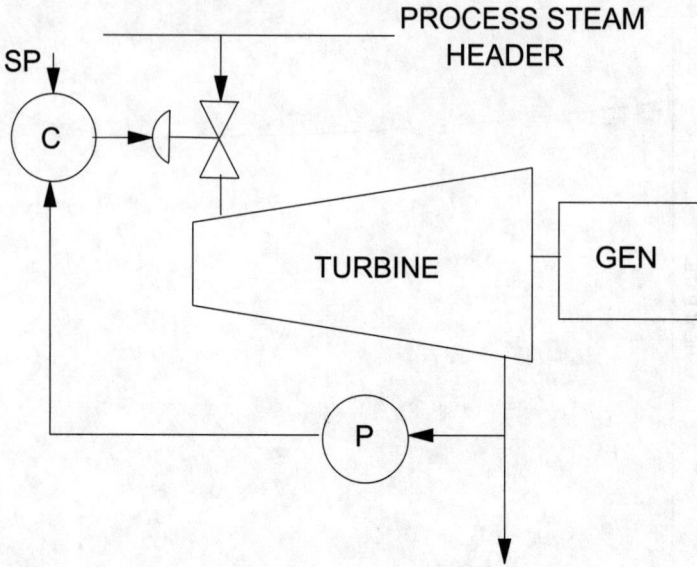

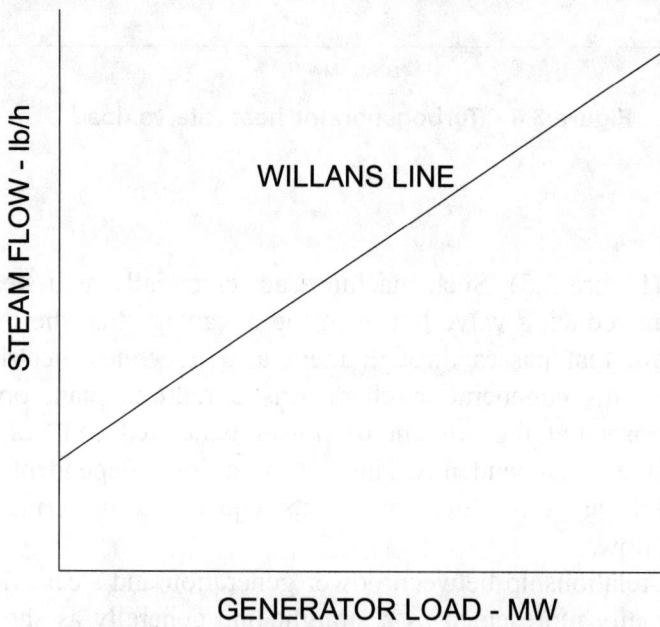

Figure 3.5   Back pressure turbogenerator

stages of a turbine. The basic equation related to energy conversion is:

Let

W_{sin} = Inlet steam flow		lb/h
W_{sout} = Outlet steam flow		lb/h
MW = Generated power		MW
a_0, a_1 = Regression coefficients		

$$W_{sin} = a_0 + a_1 MW \tag{3.19}$$

While

$$W_{sin} = W_{sout} \tag{3.20}$$

3.4.5 Turbogenerators Controlled from Surplus Steam

These machines are provided with a governor that senses the pressure at the turbine throttle valve and regulates the flow of steam through the turbine to maintain the inlet pressure sensibly constant (Figure 3.6). These machines have all the advantages and disadvantages of back pressure machines, equations (3.19) and (3.20) representing the energy conversion and mass balance relationships for these machines as well.

3.4.6 Extraction/Condensing Steam Turbogenerators

An extraction/condensing steam turbogenerator has at least two major stages and often three. As shown in Figure 3.7, live steam is supplied to the throttle of the turbine (point A), the throttle valve being controlled by the machine governor so as to maintain either speed (isochronous governor) or MW load (See also Chapter 7.0). A high pressure extraction governor valve is located part way down the turbine and, given the

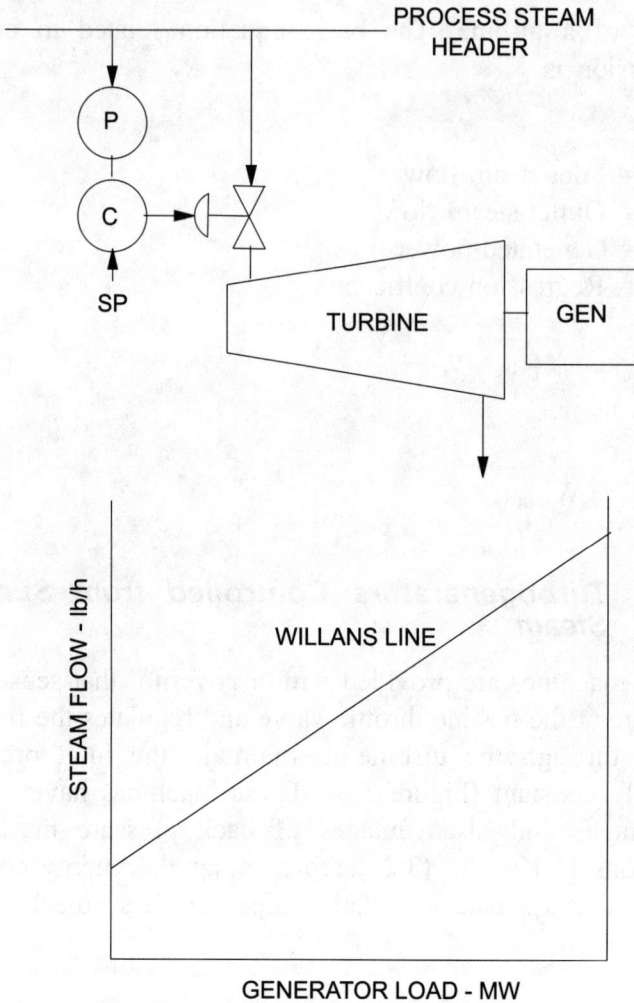

Figure 3.6   **Surplus turbogenerator**

pressure setting and throttle flow rate, allows more or less steam to pass further down the machine so a to maintain the pressure in the high pressure extraction header (point B) at the desired value. Similarly if a low pressure extraction governor is provided, it operates in a similar manner, allowing more or less low pressure steam to pass still further down the machine

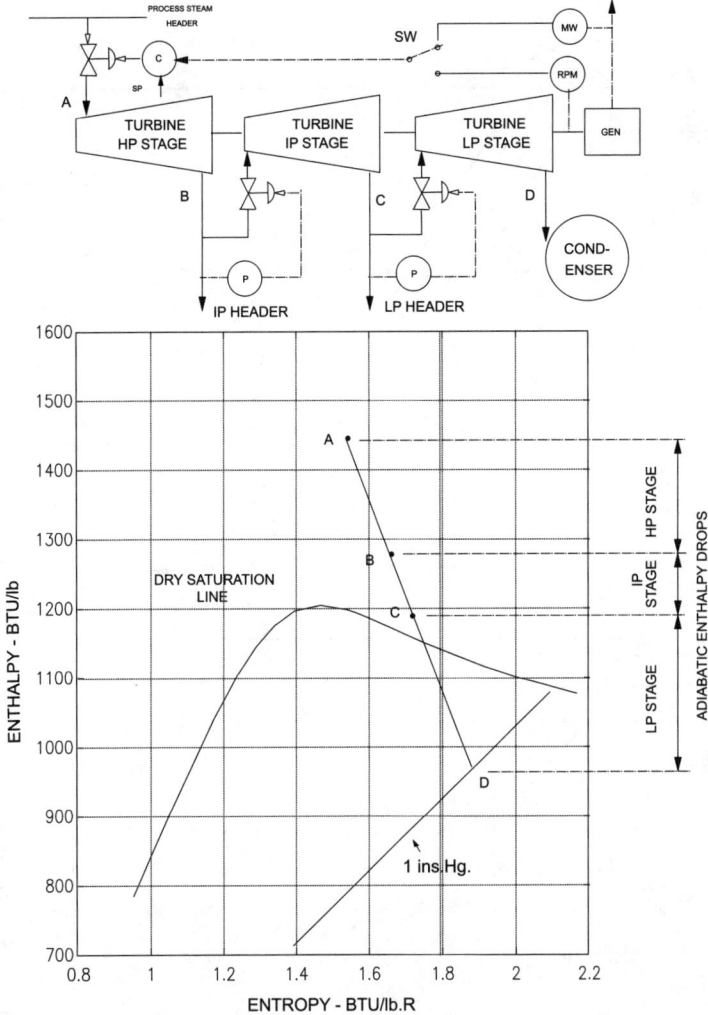

Figure 3.7 Multi-extraction/condensing steam turbogenerator

and then exhaust into the condenser (point D), so maintaining the pressure in the low pressure header (point C) at the desired value.

In the course of expanding through each stage of the turbine, the associated drop in steam enthalpy is converted into mechanical work and applies a torque to the turbine shaft that is

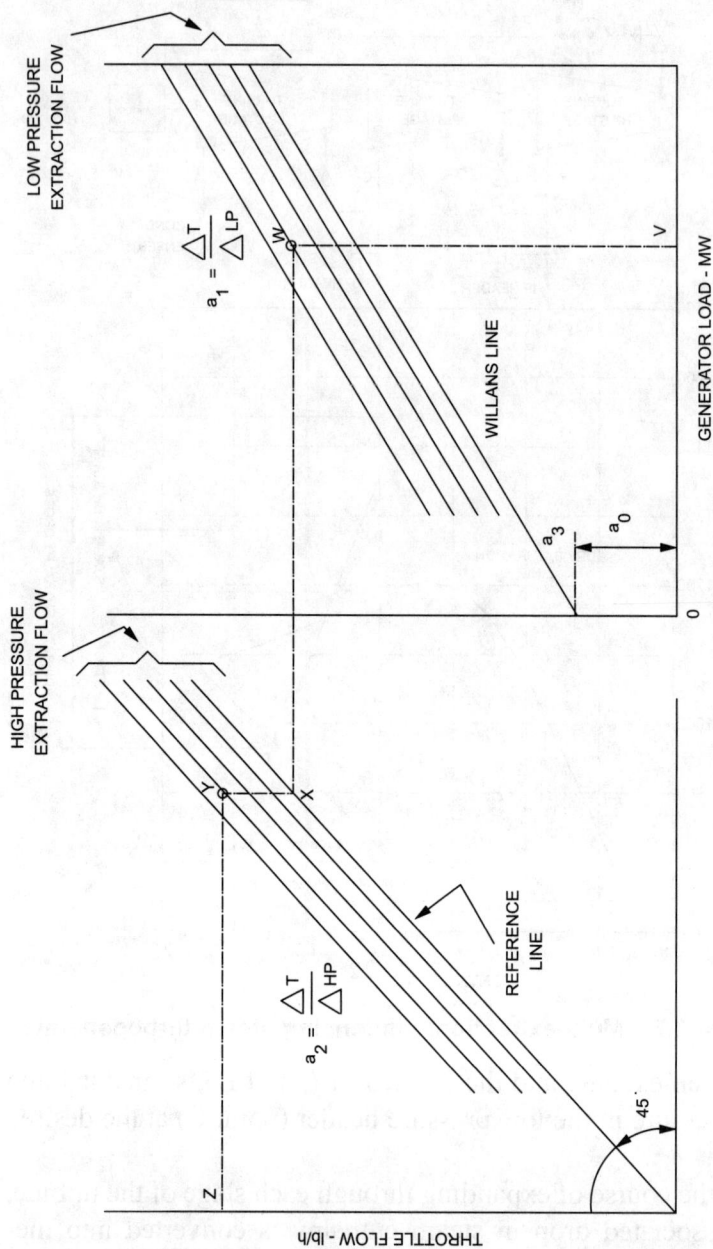

Figure 3.8   Double-extraction/condensing steam turbogenerator characteristic curve

then converted through the generator into power. Examining the expansion line in the lower part of Figure 3.7, the pressure drop across each stage is governed either by the pressure settings of the extraction governors, the live steam pressure or the condenser back pressure. For a given set of pressures and, therefore, pressure drops the adiabatic heat drop across each stage is also constant. Thus, given a sensibly constant efficiency for each stage, there will be a linear relationship between the amount of power generated (a function of the adiabatic heat drop per lb. of steam) and the steam flow through the stage.

A typical plot of the characteristic curve for a double-extraction/condensing steam turbogenerator is shown in Figure 3.8. The right-hand side of this diagram shows the low pressure stage and the Willans line. As the low pressure extraction flow rises there is a proportional increase in the flow to the throttle if the power is to be held constant, the increase in throttle flow being less than that in the extraction flow rate. Thus the right-hand side of the diagram shows how to determine the throttle flow that corresponds to a given combination of low pressure extraction flow and power.

The left-hand side of the diagram shows the effect of changes in the high pressure extraction flow on the throttle flow. Again, a change in high pressure extraction flow causes a smaller change in the throttle flow to maintain a given amount of generated power. Because of the sensibly constant adiabatic heat drops per stage as outlined above, the relationships shown in Figure 3.8 are all linear and may be expressed in the form:

$$W_{throttle} = a_0 + a_1 * W_{HP} + a_2 * W_{LP} + a_3 * MW \qquad (3.21)$$

and

$$MW = (W_{throttle} - a_0 - a_1 W_{HP} - a_2 W_{LP})/a_3 \qquad (3.22)$$

where a_0 = Intercept of Willans line on zero ordinate
        $a_1 = \Delta W_{HP}/\Delta W_{throttle}$
        $a_2 = \Delta W_{LP}/\Delta W_{throttle}$
        a_3 = Slope of Willans line

The plot of Equation (3.21) is line VWXYZ shown in Figure 3.8. Starting at point V, the generated power, the line is projected vertically upwards to point W corresponding to the low pressure extraction flow rate. The line is then projected horizontally from point W to point X on the Reference Line, which is plotted at 45 degrees to the axis of the diagram. It is then projected upwards to point Y corresponding to the high pressure extraction flow rate; finally being projected horizontally to point Z, representing the total throttle flow rate that satisfies the combined effects of the high and low extraction flow rates together with the generated power.

There are a number of major constraints on the operation of an extracting/condensing steam turbogenerator, and these may be identified as flows:

- Maximum allowable throttle flow
- Maximum load that can be provided by the generator
- Minimum load at which the generator can operate stably
- Maximum high pressure extraction flow
- Maximum low pressure extraction flow
- Minimum exhaust flow rate into the condenser
- Constraints embodied in the reactance capability curve for the generator
- Maximum condenser back pressure

Clearly, the assignments made to the steam flows and power should be such that, for a given machine, none of these constraints is violated: although a machine may be allowed to operate exactly at one or more of the above constraints.

3.5 Steam Surface Condensers

The basic principles embodied in the design of steam surface condensers have been thoroughly explored by Putman (2001). A condenser is used to capture the exhaust vapor from a steam turbine and return it to the system in the form of condensate, thus reducing the amount of make-up water that has to be treated. However, since the conditions surrounding the condenser determine the back pressure at which the vapor is condensed, it can have an important effect on the performance of the steam turbine itself. Referring to Figure 3.7, which shows an enthalpy/entropy plot of low pressure expansion line AD for a steam turbine, the lower the back pressure the lower the exhaust enthalpy. This means that more of the available energy in the vapor is converted to mechanical work. This is normally reflected in the amount of power generated and will also result in a reduced turbine heat rate.

Figure 3.9 shows a general layout of a single pass condenser, which is configured as a cross-flow heat exchanger. The vapor from the exhaust of the turbine enters at the top of the condenser shell that contains a bundle of tubes through which a supply of cooling water is pumped. There are two important equations. Let:

T_{in} = Cooling water inlet temperature   °F
T_{out} = Cooling water outlet temperature   °F
T_{vap} = Temperature of exhaust vapor   °F
W = Cooling water flow rate   lb/h
C_p = Specific heat of cooling water
Q = Condenser duty   BTU/h
A = Tube surface area   ft^2
U = Effective tube heat transfer coefficient BTU/(ft^2.h.°F)

Then

$$Q = WC_p(T_{out}-T_{in}) \tag{3.23}$$

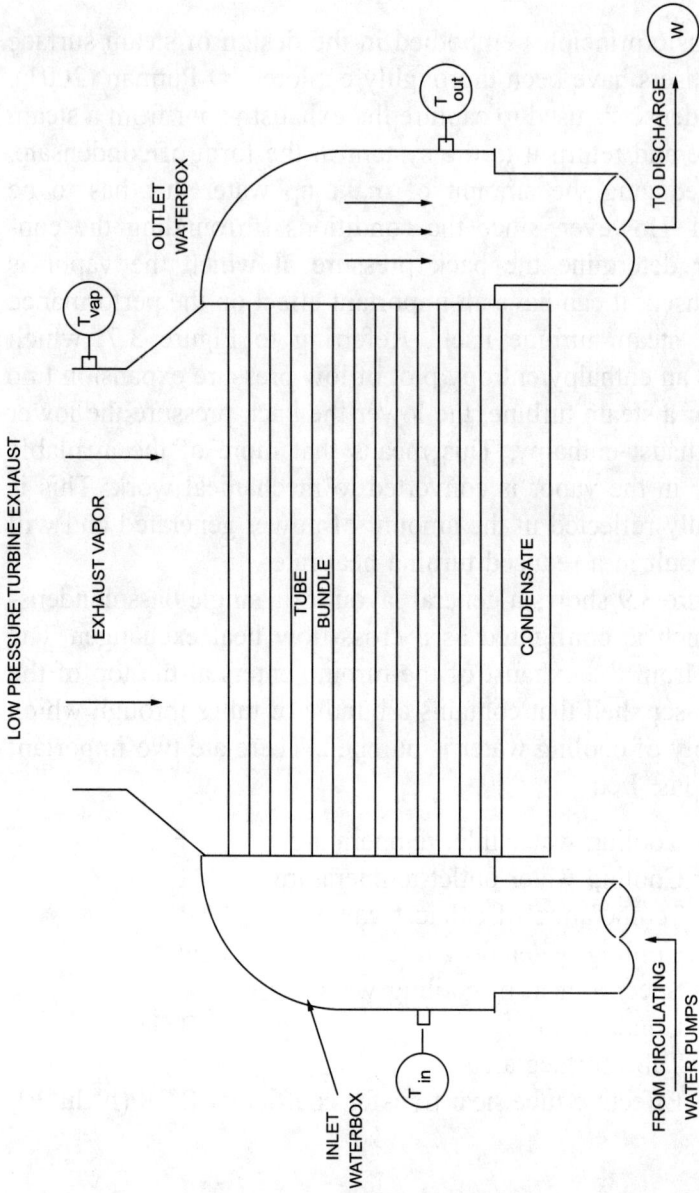

Figure 3.9 Latout of typical single-pass condenser

LOW PRESSURE TURBINE EXHAUST

EXHAUST VAPOR

OUTLET WATERBOX

T_{out}

TO DISCHARGE

W

T_{vap}

TUBE BUNDLE

CONDENSATE

INLET WATERBOX

T_{in}

FROM CIRCULATING WATER PUMPS

And

$$Q = U * A * LMTD \qquad (3.24)$$

Where

$$LMTD = \frac{T_{out} - T_{in}}{\log \frac{T_{vap} - T_{in}}{T_{vap} - T_{out}}} \qquad (3.25)$$

Unfortunately, the value of the heat transfer coefficient U will be reduced if the insides of the tubes become fouled from the precipitation of salts or silt contained in the cooling water. The value of U will also be reduced if air or other non-condensibles are allowed to accumulate in the condenser shell. For a given value of U, the condenser back pressure for a particular turbine load will be largely determined by the magnitudes of the cooling water inlet temperature and flow rate.

The maximum condenser duty should not exceed its design value nor should it be less than about 10% of that value. When the condenser duty is low, some plants try to save electrical energy by switching off pumps and so reducing the cooling water flow rate. This can be successful with a clean source of cooling water. However, if the cooling water is drawn from a lake or river and contains silt, a low cooling water flow rate can cause silt to accumulate, so reducing the U coefficient to the point where the fouling may become a maintenance concern, an appropriate plan for fouling control or mitigation should be developed. Additionally, the following sections 3.5.1 through 3.5.5 address some important maintenance and operations issues as identified by Putman (2001) to be particular to the steam surface condenser.

3.5.1 Steam Surface Condenser Operations

The basic principles of efficient operations of steam surface condensers in fossil fueled and nuclear power generating plants have been introduced by Putman (2001). In his first book titled, *Steam Surface Condensers: Basic Principles, Performance Monitoring and Maintenance,* published ASME Press, New York, NY, Putman summarizes years of successful experience in operating electric generating power plants efficiently and profitably with reduced emissions and less pollution. The critical factors impacting the steam surface condenser efficient operations include tube material selection, cooling water sources and fouling characteristics.

3.5.2 Steam Surface Condenser Maintenance

Basic successful maintenance practices of steam surface condensers in fossil fueled and nuclear power generating plants include:

- Condenser tube cleaning practices
- Leak detection
- Eddy Current testing
- Deposit analysis
- Tube plugging
- Heat Transfer Testing

All of these practices are critical. Clean condenser tubes deliver efficiency improvements (more MW generated), longer tube life, less emissions and less thermal pollution. Additional maintenance activities yield similar results.

3.5.3 Steam Surface Condenser Tube Cleaning

"Mechanical cleaners offer the most effective off-line tube cleaning method. Strong enough to remove hard deposits, they can cut the peaks surrounding pits and flush out the residue at the same time, thus retarding underdeposit corrosion" Putman (2001).

Putman (2001) further describes the mechanical cleaning process and the various types of mechanical cleaners.

"Mechanical cleaners travel through the tubes at a velocity of 10 to 20 ft/s and are propelled by water delivered at 300 psig. Some of the members of a well-known family of tube cleaners have the following features:

- *C4S.* The C4S cleaner is a general-purpose type and may be used to remove all types of obstructions, deposit corrosion, and pitting.
- *C3S.* The C3S cleaner is designed for heavy duty and is very effective in removing all kinds of tenacious deposits. Its reinforced construction also allows it to remove hard deposits, corrosion deposits, and obstructions.
- *C2X, C3X.* These types of cleaner consist of two or three hexagonal-bladed cleaner elements, having six arcs of contact per blade. They are effective on all types of deposits but are especially suitable for removing the thin tenacious deposits of iron, manganese, or silica found on either stainless steel, titanium, or copper-based tube material.
- *C4SS.* While the C4SS stainless steel cleaners can be used with all types of tube materials, they were originally developed for applications using Al-6XN stainless steel condenser tubes but have also been used for cleaning tubes in highly corrosive environments.
- *CB.* This tube cleaner was specifically developed to remove hard calcium carbonate deposits and is designed to break the eggshell-like crystalline form characteristics of these tube scaling deposits. This cleaner has been found exceptionally helpful in avoiding the need for alternative and environmentally harmful chemical cleaning methods, which were all that were previously available for removing these hard deposits.
- *H-Brushes.* Brushes are principally intended to remove light organic deposits such as silt or mud. They are also useful for cleaning tubes which enhanced surfaces (e.g.,

C4S

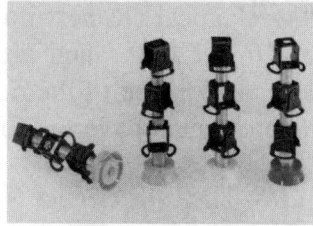

C3S

C2X

C3X

C4SS

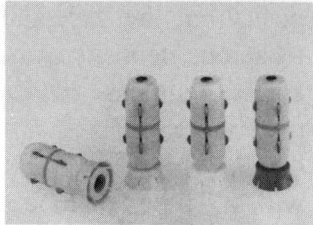

CB

H-Brush

Figure 3.10   Mechanical tube cleansers

spirally indented or finned), or those with thin-wall metal inserts or tube coatings. The brush length can be increased for more effective removal of lighter deposits."

The mechanical tube cleaners are shown in Figure 3.10.

3.5.4 Fouling Deposit Characteristics

Putman (2001) listed some of the most frequent causes of fouling:

- Sedimentary fouling or silt formation
- Deposit of organic or inorganic salts
- Microbiological fouling
- Macrobiological fouling

However, in the course of cleaning condensers, not only in the United States but also in most industrialized nations, at least 1000 different types of fouling deposit have been encountered. Each power plant has its own fouling idiosyncrasies, and even in one plant, it is not uncommon to find that the fouling characteristics are different for each unit, even when the condenser tubes are of the same ma-terial and the equipment was built to the same set of dra-wings. It is also not uncommon to find that fouling of the bundles made from copper allow tubes is of a different nature from fouling in the stainless steel bundles which are frequently associated with the air-removal section in the same condensers.

After retubing a condenser, it is also quite normal for the fouling characteristics to be significantly altered. For instance, the types of deposit encountered with stainless steel tubes (especially the manganese problem) are not the same as those found on the inside surfaces of admiralty tubes in the same location. Should a condenser be returned with titanium tubes, and then the fouling and corrosion problem will again be different from those previously experienced.

3.5.5 Deposit Sampling

From Putman (2001) we also learn that analysis of the deposits removed from the main condenser is an important part of selecting the proper cleaning process. When access to the condenser is possible during reduced-load operation or a unit outage, deposit samples can be obtained from the whole length of a tube as well as from tubes in several areas of the tubesheet. This is important in that fouling is seldom distributed uniformly throughout the condenser, because of flow and temperature variations. Thus, it is important to have a sampling plan in place for use when access to the condenser becomes possible. It has also been found that, when using water at 300 psig to propel the cleaners, deposit samples can be collected from all tube locations within any waterbox, without damage to equipment or danger to personnel.

The dry weight per unit area of the deposits collected from a tube and their composition, as determined by X-ray fluorescent analysis, can be correlated with changes in condenser performance based on historical operating data. The distribution of deposit intensity within tubes and across the tubesheet or waterbox can also assist in developing an appropriate cleaning strategy, including cleaning frequency or water treatment procedure. If tube samples in the fouled condition can be made available for heat transfer testing, the expected improvement in heat transfer from a selected cleaning procedure, or a set of alternative procedures, can also help in providing a quantifiable prediction of return on investment.

3.6   Cooling Towers

Cooling towers are to be found in many process plants, a typical configuration being shown in Figure 3.11. The cooling water circulated to the plant is returned to the hot sump, from which it is pumped to the top of a bank of cooling tower cells; each is packed with fill, down which the water to be cooled cascades. Each cell is provided with its own fan, which may be

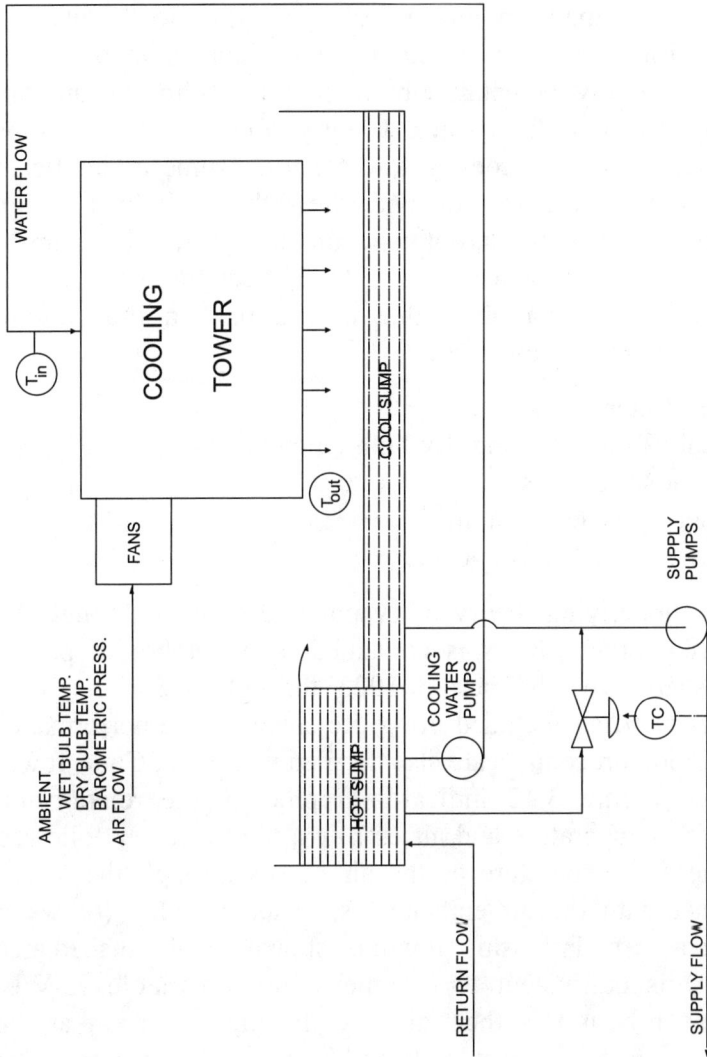

Figure 3.11   Cooling tower configuration industrial plant

run at more than one speed or may even be of the variable speed type. The fans are arranged so that the air is blown across the cascading water and is thus cooled through evaporation. The main cooling water supply flow to the plant is drawn from the cool sump but, in some plants, it is possible to raise the supply temperature by mixing in a controlled amount of the returned hot water. Clearly, the fans and pumps associated with the cooling tower can consume a significant amount of power so that, under less than full load design conditions, applying an optimization strategy to this group of equipment can save some energy. (See Section 5.6)

The primary variables that play a part in the cooling effectiveness of a tower are:

1. The water/air flow ratio
2. Ambient air wet and dry bulb temperatures
3. Barometric pressure
4. Inlet water temperature
5. Tower characteristic curve

The underlying theory of evaporative cooling through the use of cooling towers is contained in a number of papers published over the period 1923 through 1950, the most important being included in the References. The performance calculation procedures are also detailed in the CTI Code Tower (1990). Figure 3.12 indicates graphically the relationship between temperature and air enthalpy, the trajectory followed by the air temperature as the air passes through the tower, together with the air enthalpy vs. water enthalpy (or water temperature) relationship. It will be shown that the shaded area ABCD is equivalent to a dimensionless variable KaV/L. Its lower bound is the "air line, having its origin at the intersection of the wet bulb and water outlet temperatures; while the slope of this line is the liquid to gas mass flow ratio (L/G). Clearly, assuming that the heat lost by the water is equal to that acquired by the air and that the specific heat of water is

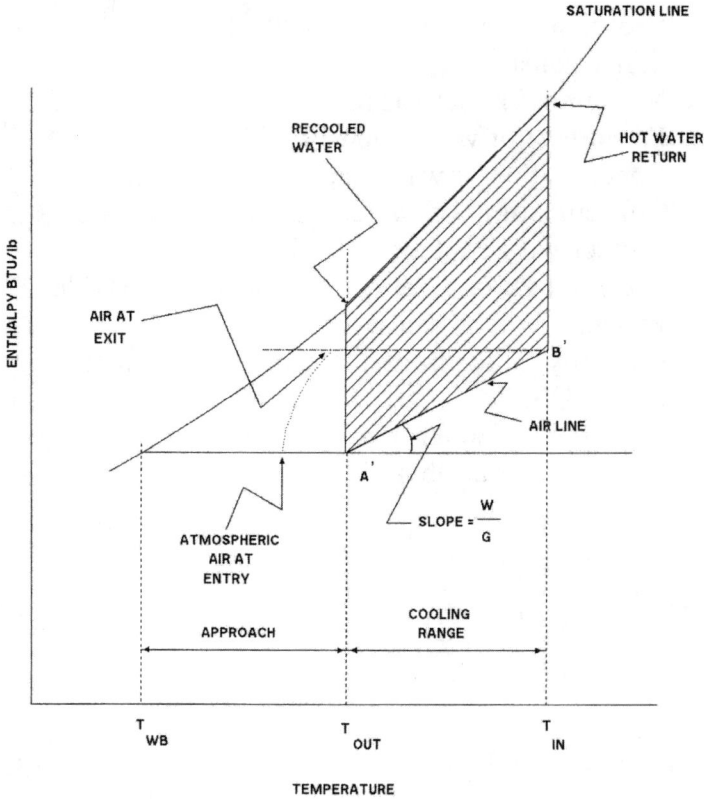

Figure 3.12   Cooling tower heat transfer diagram

unity, the change in air enthalpy divided by the change in water enthalpy (or temperature) must be equal to L/G.

3.6.1 Cooling Tower Heat Transfer Theory

The Wood and Betts (1950) presentation of cooling tower theory uses the following parameters:

H = Heat transfer rate BTU/h
T_w = Water temperature °F
T_a = Air temperature °F

K = Heat transfer coefficient BTU/(ft^2.h.°F)
A = Surface area ft^2
E = Mass transfer coefficient
H$_L$ = Latent heat of water vapor BTU/lb
x = Concentration of water vapor in ambient air lb/lb
x″ = Concentration of water vapor in saturated air in contact with cooling water lb/lb
C$_p$ = Specific heat of air at constant pressure BTU/(lb.°F)
i = Enthalpy of moist air BTU/lb
i″ = Enthalpy of saturated air in contact with cooling water BTU/lb
L = Mass flow of water lb/h
G = Mass flow of air lb/h

Then

$$dH = K(T_w - T_a) \cdot dA \tag{3.26}$$

And

$$dH = E(x' - x)H_L \cdot dA \tag{3.27}$$

Combining (3.26) and (3.27) to obtain total heat transfer rate gives:

$$dH = [K(T_w - T_a) + H_L E(x'' - x)] \cdot dA \tag{3.28}$$

From Lewis' Law:

$$E = K/C_p \tag{3.29}$$

and, substituting (3.29) within (3.28):

$$dH = K[(T_w - T_a) + (H_L/C_p)(x'' - x)] \cdot dA \tag{3.30}$$

Mollier (1923) showed that, to a close approximation, for a temperature 't':

$$i = C_p(t - 32) + x(0.46(t - 32) + 1075.4)$$

and

$$H_L = (0.46(t - 32) + 1075.4)$$

Then

$$(i'' - i) = C_p(T_w - T_a) + H_L(x'' - x)$$

Or

$$(i'' - i)/C_p = (T_w - T_a) + H_L/C_p(x'' - x) \qquad (3.31)$$

On substituting (3.31) in (3.30)

$$dH = (K/C_p)(i'' - i) \bullet dA \qquad (3.32)$$

Now

$$dH = L \bullet dT$$

thus

$$\frac{K \bullet dA}{C_p L} = -\frac{dT}{(i'' - i)}$$

from which

$$\frac{KA}{C_p L} = \int_{T_{out}}^{T_{in}} \frac{dT}{(i'' - i)} \qquad (3.33)$$

This is the shaded area ABCD shown in Figure 3.12, between the saturation line and the air line AB, and bounded by the water inlet and outlet temperatures T_{in} and T_{out}. It should be noted that KA/C_pL is dimensionless.

Assuming that the heat gained by the air is equal to the heat lost by the water, or:

$$G \cdot di = L \cdot dT$$

Then

$$(i'' - i) = (L/G)(T_{in} - T_{out}) \tag{3.34}$$

Referring to equation (3.32), the Cooling Tower Institute uses a different terminology, redefining some of the terms used above, thus:

V = Effective cooling tower volume ft^3
a = Area of cooling tower heat transfer surface Per unit of tower volume ft^2/ft^3
L = Water flow rate lb/h
K = Overall enthalpy transfer coefficient lb/(h.ft^2.lb water.lb dry air)

Then

$$\frac{KaV}{L} = \int_{Tout}^{Tin} \frac{dT}{(i'' - i)} \tag{3.35}$$

In which KaV/L is also dimensionless.

3.6.2   Cooling Tower Characteristic Curve

Figure 3.13 shows a plot on a log/log basis of the value of KaV/L vs. L/G for various approach temperatures, given the ambient air wet bulb temperature and water temperature range. The "Blue Book" published by the Cooling Tower Institute

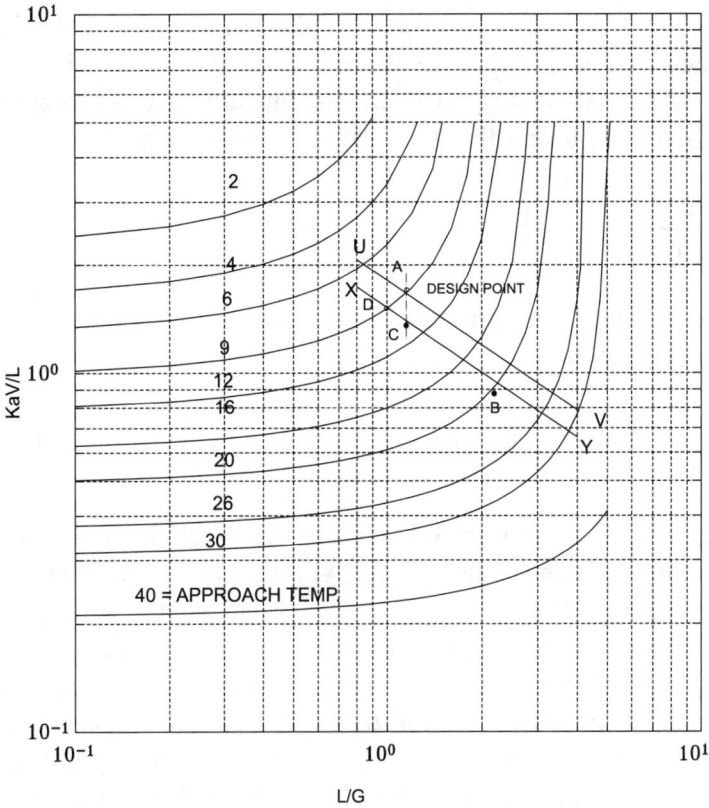

WET BULB TEMPERATURE = 75 Deg.F
RANGE = 20 Deg.F
TOWER FLOW = 300,000 GPM

Figure 3.13   Cooling tower performance curve matrix

contains a set of these curves for various values of wet bulb
temperature and range. However, a computer program can be
written incorporating the above theory to calculate and plot the
curves for selected approach values.

Lichtenstein (1943) discusses how the fundamental characteristics of cooling tower behavior might be formulated. K, aV,
L and G are the only variables that supposedly affect cooling

tower performance. If these variables can be arranged in two dimensionless groups, viz. KaV/L and L/G, then it should be found that all test points will fall on one curve. From tests, Lichtenstein found that the best relationship was:

$$\frac{KaV}{L} = c\left(\frac{L}{G}\right)^n \tag{3.36}$$

where c and n are constants. To obtain a linear plot, this can be transformed to:

$$\log\left(\frac{KaV}{L}\right) = \log(c) + n\log\left(\frac{L}{G}\right) \tag{3.37}$$

Line UV in Figure 3-13 is the characteristic curve for a cooling tower the design point (A) for which is based on a wet bulb temperature of 75°F, a range of 20°F, an approach temperature of 9°F and a design L/G value of 1.2.

Line XY represents the present characteristic curve calculated from tests. *Capability*, expressed as a percentage, is defined as the value of L/G on line XY corresponding to the design approach temperature divided by the design value of L/G. See also CTI Code Tower (1990) for further details.

3.6.3 *Predicting Cooling Tower Performance*

Switching cooling tower pumps and/or fans on and off clearly affects the amount of cooling that can be performed by a tower cell and a procedure should be available to estimate the effect of changes in the value of (L/G). Assuming that the present capability curve is known, the value of (L/G) can be used to determine the corresponding value of KaV/L. Given the present wet bulb and cooling water inlet temperatures, a search procedure can now be written to find the value of the cooling water outlet temperature that would allow equation (3.35) to be satisfied.

It should be noted in passing that, since equation (3.35) involves the integration of a reciprocal, its solution can present some computational difficulties unless certain checks are introduced. For example, the origin of the air line is determined from the wet bulb and water outlet temperatures and, in searching for the value of the latter that satisfies equation (3.35), the air line shown in Figure 3.11 must never be allowed to intersect the saturation line.

3.6.4 Air Property Algorithms

The Standards published by both the Cooling Tower Institute and ASHRAE use air property algorithms and tables based on the work of Goff and Gratch (1945) when calculating the performance of cooling towers and air conditioning equipment respectively. Unfortunately, while they used 32°F as the reference temperature for water, they used 60°F as the reference temperature for air.

Aware of this discrepancy, ASHRAE retained the National Bureau of Standards to undertake their Research Projects 216-RP and 257-RP, the purpose of which was to incorporate the most significant research on the properties of dry and moist air that had been conducted since 1945 and to provide property relationships in SI Units. The reference temperatures were to be zero °C for both air and water. The Report on these Projects was written by Wexler and Hyland (1981) and computer programs are available that allow this work to be incorporated in current engineering practice. It is not clear why ASHRAE has not yet adopted these algorithms; the delay in the widespread adoption of SI units by the engineering profession in the U.S., the cost of changing psychrometric charts, or the small improvement in the quality of air conditioning performance calculations could all have affected the decision.

However, the difference in reference temperatures could affect the values of the enthalpies that should be used when calculating heat balances in combustion processes. The more accurate air property algorithms based on the work of Wexler and Hyland, with their reconciled reference temperatures for air and water, should certainly be used in such cases.

Chapter 4
Optimization Strategies

In this Chapter, a number of optimizing strategies for industrial energy systems are detailed together with the situations in which they apply. It is possible to think of the word *optimize* as being composed of two ideas: that of *options* or choice and the idea of *maximizing* or *minimizing* the operating situation by selecting the most appropriate choice. Thus, for an optimizing situation to exist at all, it is essential that two or more choices be available to the operator, whether the choices involve different system configurations, different internal energy distributions or the availability of a number of fuels.

In many plants, the only energy in the system is that related to power. In those plants the only choice is to protect production by shedding load in the event of a system disturbance. The techniques involved in electrical load shedding are detailed in Section 9.0 and will not be discussed further here.

The optimization of a fossil-fired boiler is also outside the present discussion, except for fuel assignments when the boiler is fired with more than one fuel. Modern boiler control and optimization strategies include the use of neural networks for adaptive modeling and fuzzy logic to make appropriate decisions.

The industrial energy system to be described in this Chapter is shown in Figure 4.1. This is typical of many major industrial plants found in the pulp and paper, steel, petrochemical, refineries, manufacturing and many other industries. Such systems are closely-coupled and it is not uncommon to find that several discrete optimizing problems are nested within one

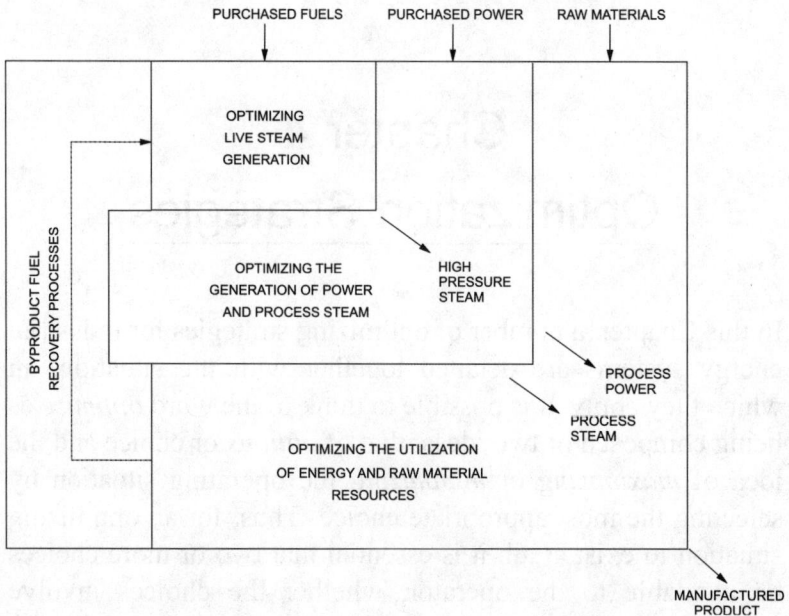

Figure 4.1   Nesting of various optimization problems within one another

another. The inputs to the system consist of fuel(s), purchased power and raw materials and Figure 4.1 shows how these are interrelated within at least three separate subsystems, each having its own unique optimal solution:

- Optimizing the *utilization* of energy and raw material resources, the task of the manufacturing departments in the plant
- Optimizing the generation of the total amount of power and process steam demanded by the plant
- Optimizing the generation of live steam

Optimizing the utilization of energy is aimed at reducing the amount of energy of all types consumed per ton of product. Good housekeeping, careful monitoring of consumption through data acquisition programs, and even the adoption of less energy-intensive processes, all fall into this category. This activity is essentially process-related and the outcome determines the quantities of process steam at different pressures, and the amounts of power and other utilities that are required to satisfy current production needs. It is important that the ability of the powerhouse to satisfy these process demands does not normally become a production constraint, but circumstances can occur (for example, on equipment outage) where the constraints become tighter. A good energy management system should be able to anticipate such situations and minimize their impact on the production processes.

Optimizing the generation of process steam and power is often a linear problem and a variety of techniques may be used for its solution. The targets used by the optimizing process are the process steam and power demanded by the production departments. It is not difficult to ensure that the optimizing functions are able to run to completion and implemented in real-time.

The contract with the local utility for the purchase of power usually contains a demand charge as well as a Kwh charge, the Kwh price often varying with the time of day usually expressed as on-peak and off-peak rates. More recently, Real Time Pricing (RTP) has been advocated by distribution utilities in which the price of energy varies each hour but not in any particular pattern. Where RTP is being applied, the optimization system must respond quickly to the changing power price, otherwise the plant may incur excessive electricity charges. While the optimizing principles do not change, there is the addition of the Information Technology environment in which the decisions must be made and implemented via a third-party interface and in real time.

On the other hand, optimizing the generation of live steam is a non-linear problem usually involving multiple boilers. However, the techniques for optimizing boiler load distribution can also be applied to other groups of non-linear equipment operating in parallel, such as chillers. The target variable for the optimized boiler load distribution is total plant steam load, this being dependent on the amount of power and process steam at different pressures demanded by the manufacturing departments, from which the optimum quantity of live steam is derived.

Figure 4.1 indicates that one element in the optimizing process is the recovery of combustible material to augment the basic fuels consumed by the boilers. In the paper industry, the efficient operation of the soda recovery boilers and the efficient combustion of bark (where this is available) can provide important economic benefits. However, recovery boilers do affect the management of the energy generation processes since the steam they produce becomes an input to the live steam balance equation. In the steel industry, the optimum utilization of blast furnace and coke oven gas is also an important consideration. Other aspects of heat recovery, e.g., the use of thermal compressors, are again process-related in that they require the presence of appropriate equipment, and so fall outside the scope of the present discussion.

The principal analytical and optimizing techniques for these kinds of closely-coupled system include:

4.1 Linear programming applications
4.2 Simplex Self-Directing EVOP
4.3 Steepest ascent methods
4.4 Equal incremental cost techniques
4.5 Optimal trajectories

4.1 Linear Programming Applications

When Dantzig (1963) published his classic text on linear programming it was used off-line as an Operations Research tool to solve engineering, business and military optimization problems. As an off-line program, it is still being used for the analysis of industrial energy systems and it was only later, with the development of microprocessors and industrial process control systems, that the linear programming algorithm came to be used for the real-time on-line optimization of those industrial energy systems to which it is applicable.

The flow sheet for a small industrial energy system that involved the cogeneration of process steam and power is shown in Figure 4.2. This power house example consists of a high pressure boiler delivering steam at 400 psig; together with a low pressure boiler delivering steam at 150 psig; a single extraction/condensing steam turbogenerator; and two steam pressure reducing valves PRV1 and PRV2, the former being equipped with a desuperheater injecting condensate drawn from the condenser into the valve discharge. The plant is also connected to a tie-line from the local utility from which power can be purchased in accordance with a contract.

The first step in the analysis of such a system is to construct the flow sheet to show the major items of equipment and all their interconnections. Note that Figure 4.2 shows only the paths followed by the various energy resources, each of which should be identified by a numbered node. In this case these resources consist of:

1. TG throttle flow
2. TG extraction flow
3. TG condenser flow
4. TG power
5. 400/150 steam reducing valve flow PRV1
6. 150/50 steam reducing valve flow PRV2
7. HP boiler flow
8. LP boiler flow
9. Tie-line power

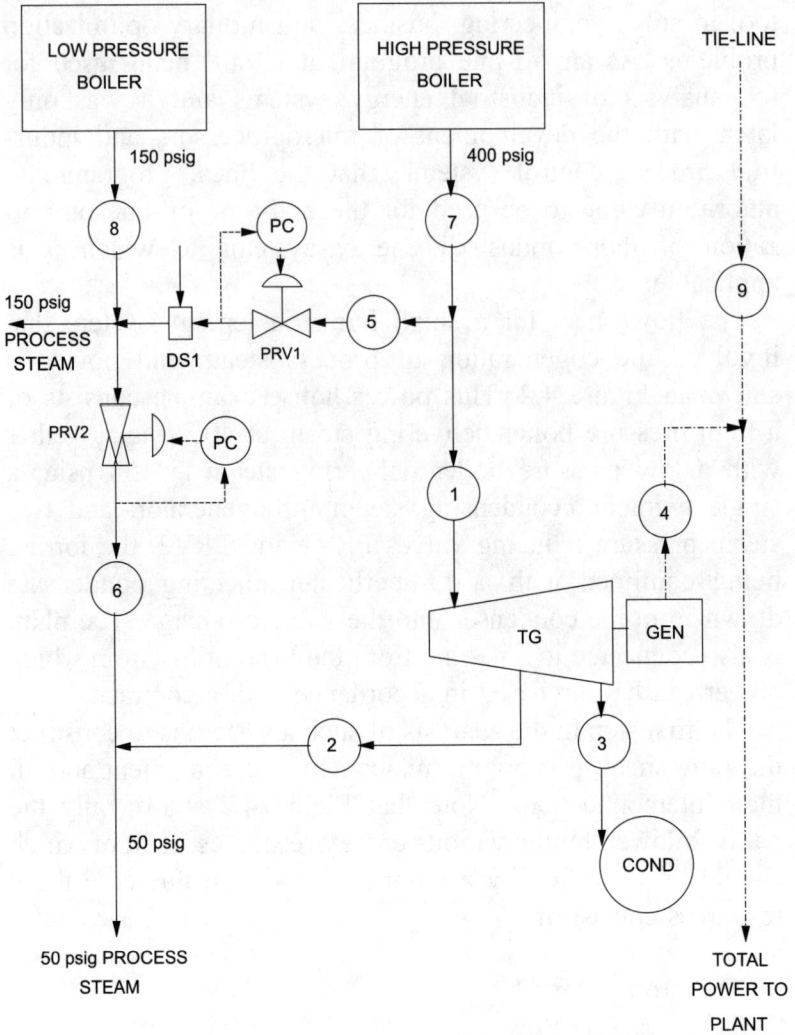

Figure 4.2   Small power house flow sheet

The first rule for a linear programming problem is that **All** variables can have only either positive or zero values in the solution. Negative values in the solution cannot occur. However, there are cases when a resource flows in both directions, e.g. where power can either be imported or exported through the tie-line. When this occurs, two separate variables must be assigned to this resource, the variables in the appropriate balance equations in the matrix being signed appropriately.

The next step is to construct the linear programming matrix for this flow sheet, the result being shown in Table 4.1. The matrix consists of the coefficients contained in a set of inequalities, there being a total of 17 in this case. A number of these inequalities are upper design or operating constraints for the resources to which they apply while other inequalities define the lower constraints that pertain to this system. The remaining inequalities (they are actually equalities) define the significant steam and power balances within the system.

Before the matrix can be defined, it is necessary to establish the envelope of operating conditions within which the variables for a machine must lie when the solution converges. The shaded area in Figure 4.3 shows a typical area and it will be noted that the boundaries can be defined either as constants or as straight lines.

Equality #5 is of interest in that the coefficient 1.13 in column #5 defines the total flow of steam delivered by pressure reducing valve PRV1 (resource #5), the fraction 0.13 being the desuperheating water/steam flow ratio (See also Section 3.3). Note that the desuperheating water flow itself is not explicitly defined as a resource, since it has been found that the fewer the resources the faster the solution converges. Equality #13 is also of interest, being the throttle flow equation for the turbogenerator, rearranged to conform to the rules for constructing a linear programming matrix. Equality #13 might also be termed the energy conversion equation and defines the interrelationship between the throttle flow to the turbine, both extraction flow(s) and power. As shown in Section 3.4, it is derived from the design data.

Table 4.1   Linear programming matrix

Equation number	Equation	Energy resources									<	Constant
		1	2	3	4	5	6	7	8	9		
1	Maximum Throttle flow	1									<=	180,000
2	Maximum Extraction Flow		1								<=	120,000
3	Maximum Condenser flow			1							<=	140,000
4	Maximum Power				1						<=	9,500
5	Maximum HP boiler flow					1					<=	200,000
6	Maximum LP boiler flow						1				<=	100,000
7	Maximum Tie-line power									1	<=	3,800
8	Turbine steam balance	1	-1	-1							=	0
9	HP header balance	-1									=	0
10	150 psig process steam balance					1.13	-1	1			=	150 psig process steam
11	50 psig process steam balance		1				1				=	50 psig process steam
12	Power balance				1					1	=	Plant power
13	Throttle flow equation	1	-0.5		-8.7						=	7,200
14	Minimum condenser flow			1							>=	10,000
15	Minimum power				1						>=	2,000
16	Minimum HP boiler flow							1			>=	20,000
17	Minimum LP boiler flow								1		>=	10,000
Cost								0.005	0.0045	0.04		

Resources:

1   TG throttle flow
2   TG extraction flow
3   TG condenser flow
4   TG power
5   400/150 reducing valve flow
6   150/50 reducing valve flow
7   HP boiler flow
8   LP boiler flow
9   Tie-line power

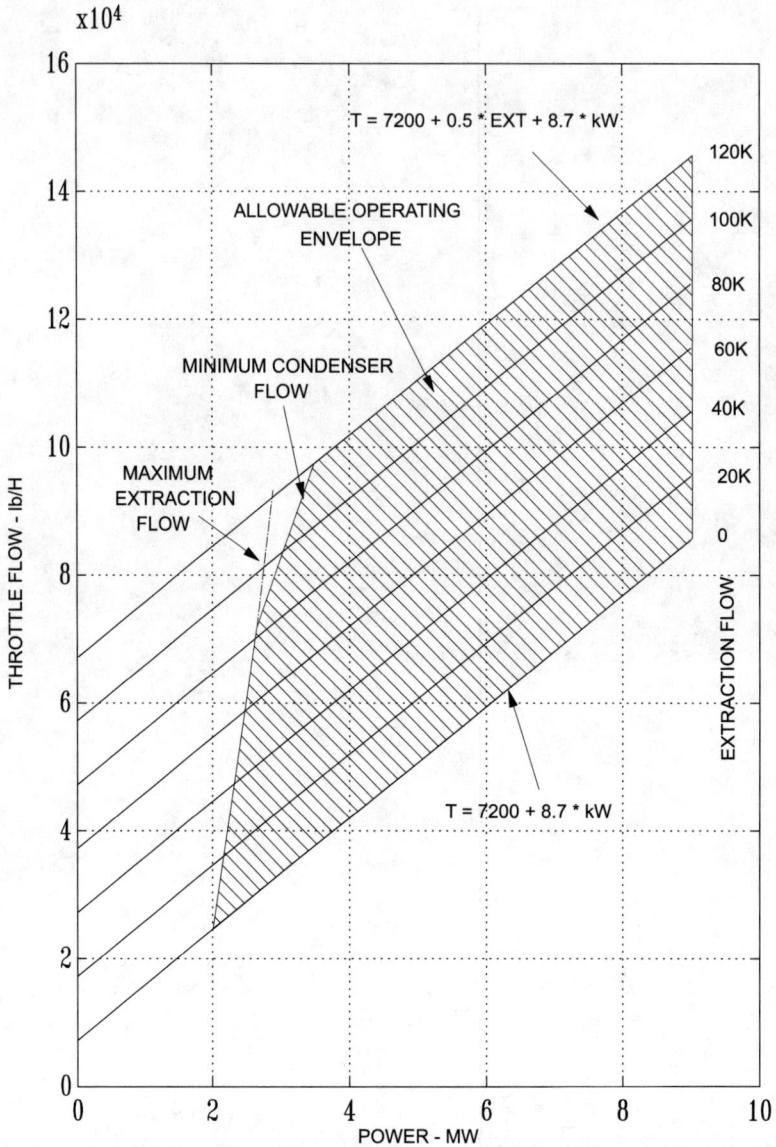

Figure 4.3 Characteristic curve for single extraction/
condensing steam turbogenerator

Meanwhile, the three demands for energy from this power-house (150# and 50# process steam and total plant power) are completely defined in the constants assigned to the right hand side of equalities 10, 11 and 12, these being updated as plant conditions and energy demands change.

The other important line in Table 4.1 is the cost function, in which the unit costs of appropriate resources are assigned to the corresponding columns in the cost function line. The convention is that a positive cost is assigned to purchased resources and negative costs to exported or revenue resources. In this case, positive unit costs are assigned only to the high pressure and low pressure steam sources because of their fuel costs, as well as the tie-line power energy cost. Where a power contract contains on-peak and off-peak prices, the value appropriate to the time must be stored in the column of the cost function assigned to power, in order to correctly perform the analysis for a given set of conditions and at a specific time of day. After execution, the optimum contains the minimum cost in purchased resources for a given set of combined energy demands. Note that the internal costs of other resources are not relevant to the calculation of the solution.

It should also be observed that there are no heat balances in this matrix. It has been found not only that they are redundant but they also complicate the matrix and increase the time taken for the program to converge. In practice, while the steam enthalpies do vary with the values of extraction and condensing flows, PRV flows, etc., the process itself will respond to such changes by automatically adjusting to the demands for steam at the various pressures.

Occasionally it is desired that an equipment item be removed from the matrix. Clearly this can be accomplished by removing the corresponding set of equalities and inequalities. However, a more convenient approach is to dedicate a new variable as a switch (See SW of Table 4.2) which can have a value of either zero or unity, all of the constants and constraints

Table 4.2 Matrix configuration with equipment switch

Equation number	1	2	3	4	SW	<	Constant
1 Maximum throttle flow	1				−180000	<=	0
2 Maximum extraction flow		1			−120000	<=	0
3 Maxdimum condenser flow			1		−140000	<=	0
4 Maximum power				1	−9500	<=	0
8 Turbine steam balance	1	−1	−1			=	0
13 Throttle flow equation	1	−0.5		−8.7	−7200	=	0
14 Minimum condenser flow			1		−10000	>=	0
15 Minimum power				1	−2000	>=	0
18 Switch control					1	=	zero or unity

previously shown in the matrix as right-hand side values now being transferred to this new column. When SW = 1, the matrix behaves exactly as the matrix of Table 4.1: but when SW is set to zero, all variables and constants associated with this set of inequalities are forced to zero. In this way, an equipment item can be completely removed from the solution merely by changing the value of a single variable SW.

4.1.1 Commentary

The results from running the linear programming matrix shown in Table 4.1 are given in Table 4.3. Three cases were evaluated: (a) the base case with all equipment functioning and plant demands of 40,000 lb/h for HP steam, 30,000 lb/h for LP steam and 7,000 KW; (b) the same with the desuperheater eliminated; (c) the same as (a) but with the LP boiler shut down. The optimizing tendencies are clearly illustrated.

In case (a), all the low pressure steam is being provided by extraction from the turbogenerator and there is no flow through PRV2. However, if the plant were to demand an increase in power, this would cause the extraction flow to be reduced and PRV2 to open.

In case (b), with the desuperheater eliminated, it is no longer profitable to pass high pressure steam through PRV1 to reduce the load on the LP boiler, which is now supplying all of the 150 psig process steam. In the plant shown in the flow sheet of Figure 4.1, the desuperheater was originally connected to the boiler feedwater system and the dissolved solids that precipitated in the ports of the desuperheater caused frequent maintenance. Not realizing its importance in reducing energy consumption, it was even taken out of service. However, subsequent analysis showed that this desuperheater would be beneficial and the desuperheater was brought back into service with the condenser condensate system, free from dissolved solids, now the water source.

Table 4.3   L.P. matrix results

	Base case	No desuperheater	No LP boiler
Plant HP steam demand	40,000	40,000	40,000
Plant LP steam demand	30,000	30,000	30,000
Plant power demand	7,000	7,000	7,000
1 TG throttle flow	50,040	50,040	50,040
2 TG extraction flow	30,000	30,000	30,000
3 TG condenser flow	20,040	20,040	20,040
4 TG power	3,200	3,200	3,200
5 400/150 reducing valve flow	26,548	0	35398
6 150/50 reducing valve flow	0	0	0
7 HP boiler flow	76,588	50,040	85,438
8 LP boiler flow	10,000	40,000	0
9 Tie-line power	3,800	3,800	3,800
Cost - $/h	580	582	579

In case (c), even after taking the LP boiler out of service, there is only a slight improvement in operating cost.

4.2   The Simplex Self-Directing Evolutionary Operation (SSDEVOP) Optimizing Technique

When G.E.P. Box (1957,1959) first presented his Evolutionary Operation (EVOP) concept it was for the marginal improvement of chemical processes, even batch processes. In his original paper he outlined an experimental design which consisted of eight experiments in which the values of a selected set of variables were perturbed in an organized pattern by small amounts. Using the process itself as the feedback, the results of each experiment were tabulated, initially on a chalk board mounted adjacent to the control panel. The cost of each experiment was also tabulated in the form of a figure of merit based on yield, unit cost of product, and/or tolerance on product specification, etc. At the end of the set of experiments, calculations were performed based on the individual figures of merit and the experimental values, the result indicating the new set of process variables that should be adopted in order to move the process to a more optimal state. This technique was widely used by the process industries and even successfully penetrated the field of mineral ore dressing (Eggert (1967)).

In 1962, Spendley et al (1962) suggested that a Simplex design be applied to the fundamentals of Evolutionary Operation and Carpenter and Sweeney (1965) described how this enhancement, which came to be known as the Simplex Self-Directing Evolutionary Operation Technique (or SSDEVOP), had been applied to a number of chemical processes. There are five steps to be taken when planning an application of SSDEVOP:

1. Select the important process variables over which control is to be exercised
2. Select the dependent variable (cost function or figure of merit) that is to be maximized or minimized
3. Define the constraints such as design constraints, product quality specifications and safety limits
4. Specify the starting or base experiment
5. Define the rules to be used to determine the set of perturbed data sets that constitute the subsequent sets of experiments

Note that SSDEVOP can be used as an optimum seeking method for both linear and nonlinear problems but, if convergence is to be achieved in real time, it is necessary to restrain the number of variables that need to be perturbed during the search.

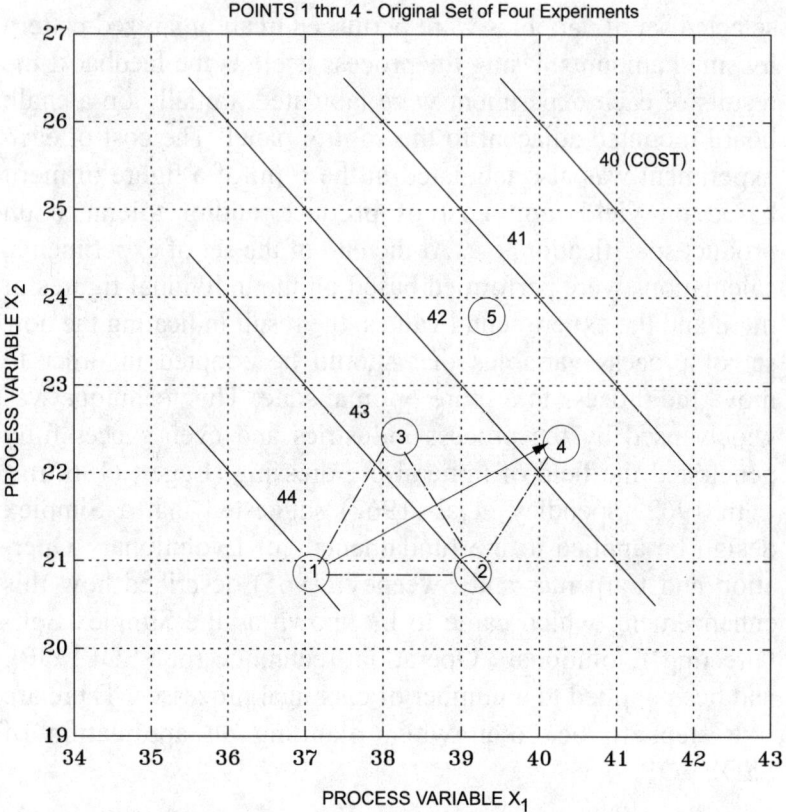

Figure 4.4   Simplex experimental design

4.2.1   The Simplex Principle

The Simplex principle is illustrated in Figure 4.4. Consider a process with two perturbable variables X1 and X2, the values of which are sought when the cost or figure of merit assumes its minimum value. The starting point is experiment #1, after which variable X1 is perturbed to define experiment #2. Experiment #3 is selected by forming an isosceles triangle having experiments #1 and #2 as its base. The conditions for

VARS.	E2	C1	C2	COST
1	E_2 BASE - a1	C_1 BASE - a2	C_2 BASE - a3	B1
2	E_2 BASE + a1	C_1 BASE - a2	C_2 BASE - a3	B2
3	E_2 BASE	C_1 BASE + 2 a2	C_2 BASE - a3	B3
4	E_2 BASE	C_1 BASE	C_2 BASE + 3 a3	B4

5	AVERAGE VALUES CASES 1, 3 and 4			
6	TWICE AVERAGE VALUES CALCULATED IN STEP 5			
7	SUBTRACT VALUES OF WORST CASE #2			NEW BASE CASE

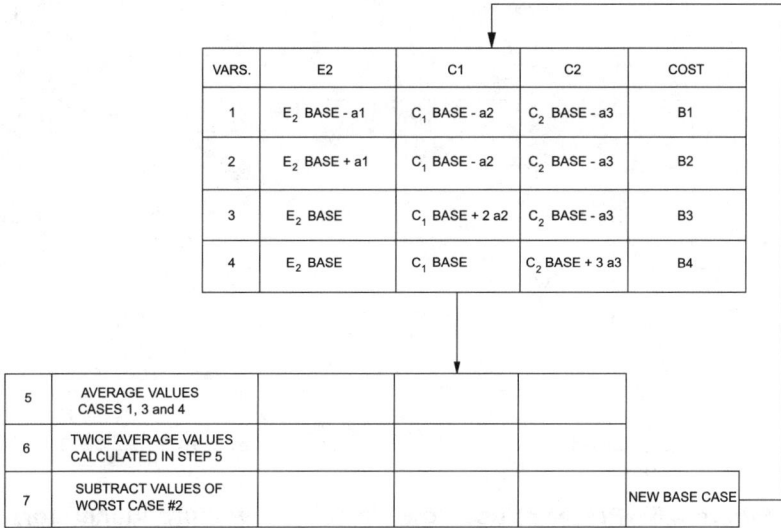

Figure 4.5   SSDEVOP experimental design

experiment #4 can be determined by pivoting triangle 123 about the base 2–3. The four experiments are then conducted and the response of each in terms of cost or figure of merit is noted.

It is found that experiment #4 is the best case, followed by experiment #3. A new experimental point (#5) may now be selected by pivoting triangle 234 about base 3–4, knowing that experiment #5 will move the process towards a more optimal point.

4.2.2   SSDEVOP Experimental Design

The form of the SSDEVOP experimental design recommended by Carpenter and Sweeney (1965) and Eggert (1967) is given in Figure 4.5. This shows the experimental tableau or experimental design for a process in which three (3) process

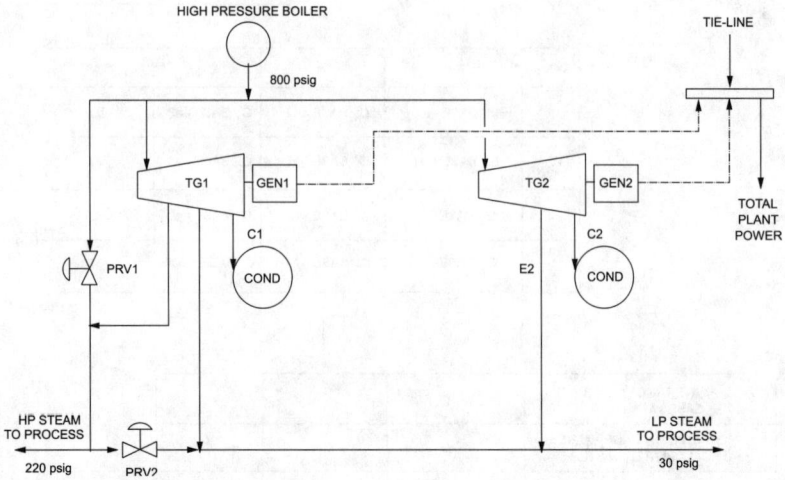

Figure 4.6   Power house flow sheet with two turbogenerators

variables can be perturbed, the flow sheet for one example being contained in Figure 4.6. The base values of the perturbable variables are respectively E2, C1 and C2, the perturbations to be applied to each being selected as a1, a2 and a3 respectively. The initial values selected for these constants are a matter of good judgment.

For experiment #1, the perturbations are all *subtracted* from the base values of E2, C1 and C2. For experiment #2, a1 is *added* to the base value of E2; while a2 and a3 are both still subtracted from their associated base values. For experiment #3, E2 assumes its base value; *twice the value of a2 is added* to the base value of C1; while a3 is still subtracted from the base value of C2. For experiment #4, E2 and C1 are both assigned their base values; while *three times the value of a3 is added* to the base value of C2. It will be noticed that the sum of the perturbations for each column or variable is zero, so avoiding any bias being introduced into the selection of the next base case. After each experiment has been implemented, the costs (B1 thru B4) are noted and added to the appropriate line in the fifth column: At the end of the set of experiments, one of them will be selected as being the worst case.

To calculate the values of E2, C1 and C2 that will constitute the new and, probably, more optimal base case that is to be used in the next set of experiments, the following steps are taken:

5. Calculate the average for each of E2, C1 and C2 from the three sets of values obtained from the three better experiments
6. Double the average value that was calculated for each variable in step 5
7. From the values in the line calculated in step 6, subtract the E2, C1 and C2 values obtained from the worst case experiment, i.e. the one that showed the highest cost.

The set of values calculated in step 7 forms the *new base case* and the process can continue ad infinitum or until no further cost improvements are shown, so that this set of operating conditions can be assumed to be the optimum. Penalties are applied to the calculated cost whenever some variable encounters a constraint, thus avoiding optimal solutions that cannot be implemented in practice. Alternatively, if a variable continues to encounter the same constraint, it can be removed from the experimental design set. Note that by reducing the values of a1, a2 and/or a3, a possibly more accurate result may be obtained.

4.2.3 SSDEVOP Optimization of a System with Two Turbogenerators

As an example of a three-dimensional linear problem amenable to solving using the SSDEVOP approach, Smith and Putman

(1983) discussed the flow sheet shown in Figure 4.6. This power house consisted of a range of boilers, two extraction/condensing turbogenerators TG1 and TG2 and a tie-line. The high pressure extraction from TG1 supplied 220 psig steam to the process and was augmented by passing live steam from the boilers through pressure reducing valve PRV1. Both turbines extracted process steam at 30 psig, this also being augmented by passing 220 psig steam through pressure reducing valve PRV2 as required. Assume that, as indicated in Section 3.3, both machines exhibit linear relationships between throttle flow, extraction flow(s) and power, and that the demands for process steam at the different pressures and power are all givens.

It will be observed that, although there are at least 13 energy resource flows in this system, in this case it is only necessary to perturb the extraction flow E2 on one machine and the condenser flows C1 and C2 on both machines. The key lies in the turbogenerator throttle flow equation discussed earlier in Section 3.3. Let:

$$T = \text{Throttle flow}$$
$$HPE = \text{High Pressure Extraction flow}$$
$$LPE = \text{Low Pressure Extraction flow}$$
$$C = \text{Condenser flow}$$
$$MW = \text{Power}$$

c_0, c_1 c_2 and c_3 = equation constants (possibly obtained by regression analysis of data)

Then

$$T = c_0 + c_1 * HPE + c_2 * LPE + c_3 * MW \qquad (4.1)$$

And

$$T = HPE + LPE + C \qquad (4.2)$$

While

$$MW = (T - c0 - c1 * HPE - c2 * LPE)/c3 \qquad (4.3)$$

Equations (4.1–4.3) apply directly to the double-extraction machine TG1. For TG2, the value of HPE will be zero.

The power generated by each turbogenerator can be calculated from knowledge of the steam flows: and by subtracting the total generated power from the total plant power demand, the tie-line power can be obtained. Thus, the use of only three variables completely specifies the generated power, tie-line power and throttle flows throughout the system as well as the cost of running the plant under the stated set of conditions.

The method has one important advantage over linear programming in that solutions are not always forced toward constraints, but will often lie well inside the allowable operating ranges. This is an important consideration when there is a reversal, for example, of a generate/buy decision, as in the transition from an on-peak to an off-peak period. With linear programming, the optimal set of assignments can be markedly different if reversal occurs and they can oscillate unless care is exercised. But with the SSDEVOP method, the transition is much more graceful and less extreme.

It is of interest to note that, when the flow sheet of Figure 4.6 was evaluated using SSDEVOP, the results were exactly the same as if this linear problem had been solved using linear programming.

The SSDEVOP procedure has been used not only for optimizing a system with two extraction/condensing steam turbogenerators but also for such diverse problems as: optimizing the load and fuel distribution among two or three boilers each fired with multiple fuels; for optimizing the distribution of excitation among generators and synchronous motors so as to minimize internal electrical system power transmission costs; and for optimizing the proportional, integral and derivative settings of a single loop process controller.

4.3   Steepest Ascent Method

Difficulties in analyzing optimization problems sometimes occur if the topology or shape of the response surface generated by the perturbation of variables is not known with certainty. Steps can be taken to explore this topography by surveying the mathematical model of the process within the envelope containing the variables of interest. Peaks and valleys will be identified and judgments made or rules developed to locate the optimum surface, minimum or maximum, for the system. Because of the topological uncertainties, where search strategies starting from a selected point are employed, sub-optimal points may be falsely identified as the optimum. This is rather like identifying as the optimum a dimple pressed into the side of a response surface otherwise represented as a smooth bowl.

Hooke and Jeeves (1961) studied this problem and derived the steepest ascent method as a means of accelerating the search; but, in the case of an overshoot, allowing the search to return to a previously established best case, so allowing the search to be pursued in a different direction.

Wilde (1964) gives an excellent review of the method, which is illustrated in Figure 4.7 for a two-variable (x_1, x_2) maximizing problem. A starting point (A) is picked and perturbations made in several directions to establish the response. From among this initial set, that direction which has the slope with the steepest ascent is chosen for the next step. The two variables are then perturbed in this direction with a vector of length z_1, so as to achieve point B and a new multi-directional

X2

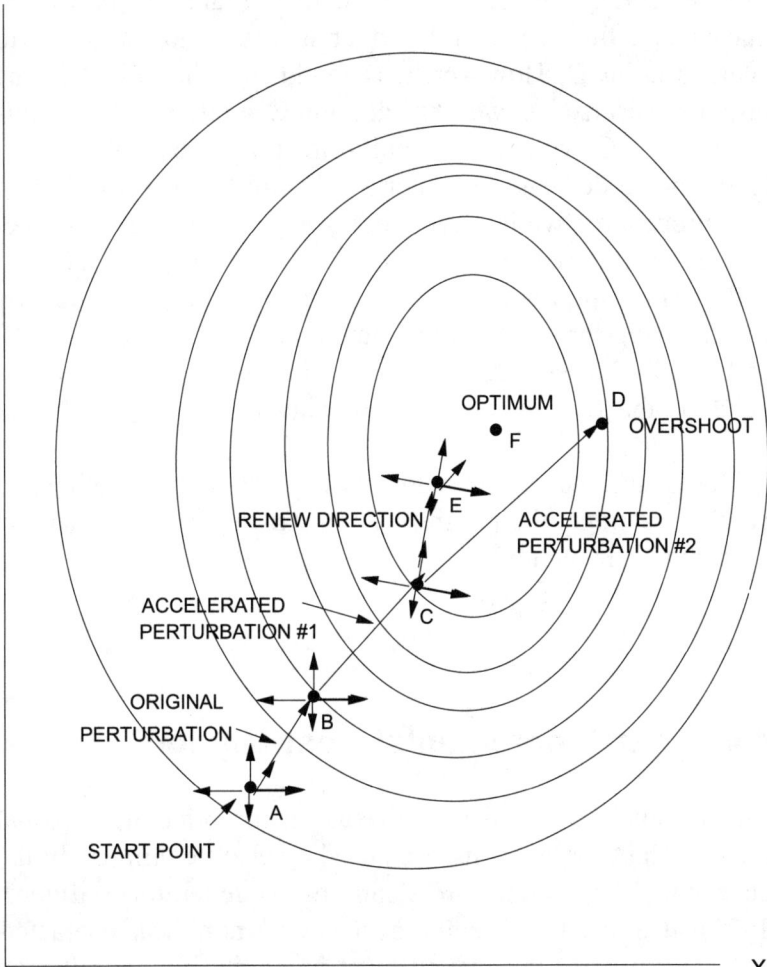

Figure 4.7   Steepest ascent mapping

search is conducted. Once the new direction of steepest ascent has been selected, the search is accelerated by increasing the length of the vector from z_1 to z_2, so as to achieve point C and an improvement is verified. Assuming an improvement has been detected, another multidirectional search is now conducted and the vector still further increased to z3, so as to achieve point D. However, it is found that the height of the response surface is lower than at point C and the search begins again from C with a new multi-directional scan, the vector length being decreased to its original value of z1. By repeating this procedure, while even decreasing z1 to obtain a finer resolution, the true optimum at point F is found.

That the optimum has been found can be confirmed by commencing the search from some other point on the map. If, at the end of the search, the optimum has the same location and value as the first search, it is probably the true optimum for that problem.

Hooke and Jeeves also studied the effect on the search when climbing a ridge, the progress along which can be accelerated by using their method.

4.4   Equal Incremental Cost Method

This optimal search method, based on the principle of equal incremental cost, has long been used by electric utilities for the economic dispatching of steam turbo-generators (Elgerd, 1971). It is used with equipment having nonlinear characteristics, a category into which most industrial boilers fall. The technique assumes that all units in the optimizing set discharge into a common header (steam) or are connected to a common bus (power). In the case of boilers, it is assumed that only one kind of fuel, or a fixed fuel mix, is being fired in each boiler so that a continuous efficiency vs. load curve can be generated.

Elgerd describes the classical method for solving this problem in the utility industry, in which a LaGrange multiplier in a set of partial differential equations is adjusted until the

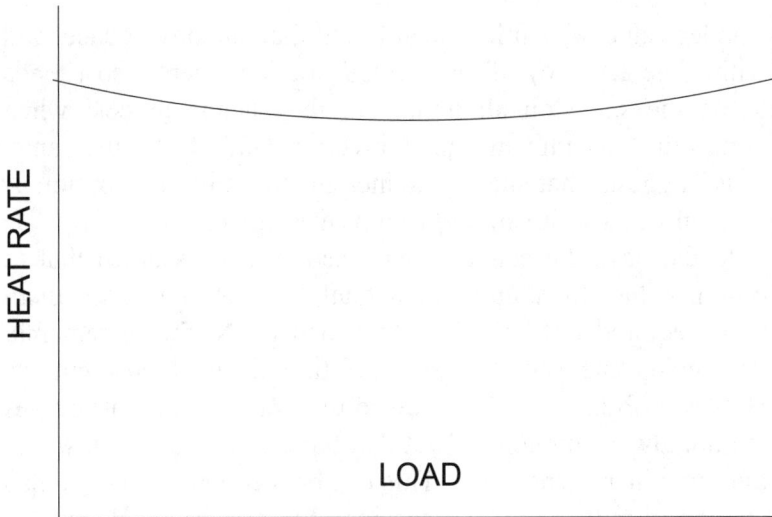

HEAT RATE vs. LOAD

Figure 4.8

total desired load is satisfied, with the same incremental cost being exhibited by each unit. Our approach to the problem was from first principles. Assuming N units in the optimizing set, each with its own cost vs. load (i.e. characteristic) curve, and a total load L, then the initial load assigned to each unit was L/N. The total cost was then calculated, together with the change in cost on each unit if its load were to be raised by x MW and also if its load were to be reduced by the same amount. Differences in the characteristic curves made these cost changes different for each unit. The program then identifies that unit which will provide the greatest reduction in total cost if its load is lowered; as well as that other unit which will provide the least increase in total cost if its load is raised. The total load is then re-distributed and a new total cost calculated. This process is repeated until no improvement in total cost is found. If greater accuracy is required, the search process may be repeated with a smaller value of x. It has been found that, at convergence, and within the accuracy of the method, the incremental cost tends to be the same on all units, i.e. the change in cost when perturbing any unit in either direction tends to be the same. This suggests that the equal incremental cost distribution is optimal because it cannot be improved upon.

In the past, the conventional wisdom had assumed that to optimize the distribution of a total load among N *identical* units required only dividing that load by N, the incremental cost being the same because of the identical cost curves. However, Putman (1978) showed that incremental cost curves are not always monotonic but that there can be two loads with the same incremental cost. This can be a complication but also an opportunity, leading occasionally to a reduced cost of operation if the lower of the two loads is selected for inclusion in the optimal distribution (See Figure 4.10)

Figure 4.8 shows a typical efficiency vs. load curve, while Figure 4.9 indicates the corresponding heat input vs. load curve, the heat input being the product of load and heat rate divided by boiler efficiency. The incremental cost plotted in Figure 4.10 is a plot of the differential coefficient of the curve of Figure 4.9 and shows that, within the total load range, there may be two loads, A and B, having the same incremental cost.

To investigate this further, experience has shown that the heat rate vs. load curve for a boiler/turbogenerator unit can be described as a second order polynomial of the form:

$$H = a_0 - a_1 P + a_2 P^2 \qquad (4.4)$$

Where

H = Heat rate   BTU/kwh
P = Power   Kw
a_0, a_1 and a_2 regression coefficients

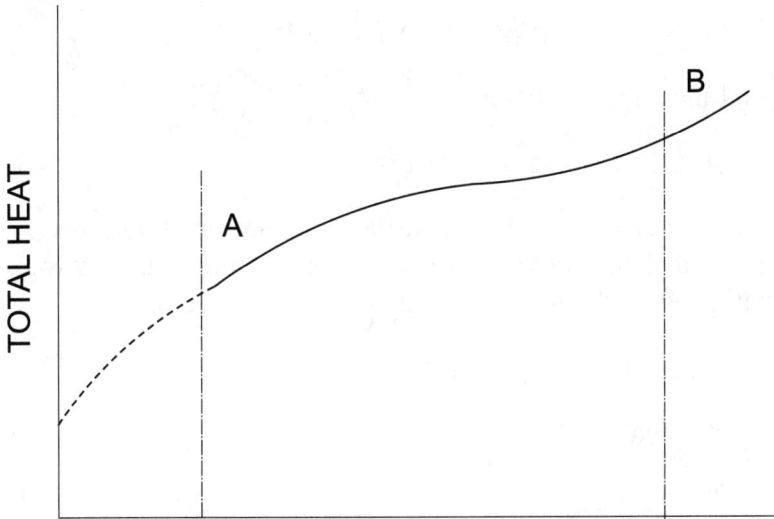

TOTAL HEAT vs. LOAD

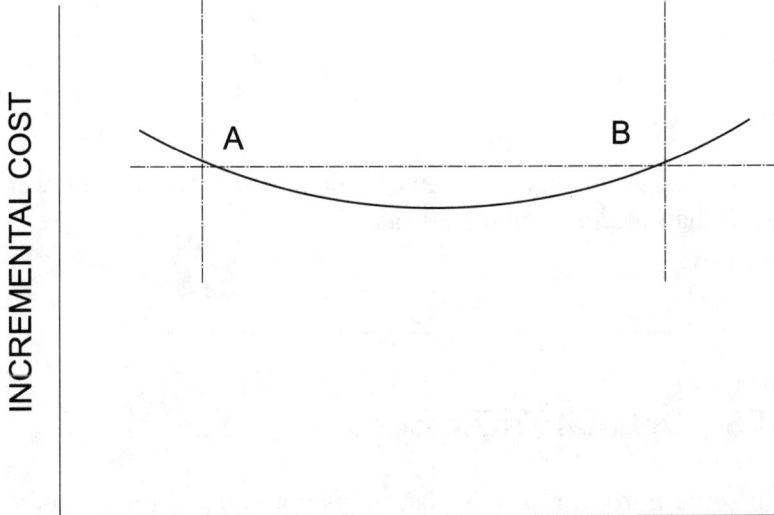

INCREMENTAL COST vs. LOAD

Figures 4.9 and 4.10

$$\text{Cost} = C = HP = a_0P - a_1P^2 + a_2P^3 \qquad (4.5)$$

And incremental cost is:

$$\partial C/\partial P = a_0 - 2a_1P + 3a_2P^2 \qquad (4.6)$$

Clearly, for a given incremental cost, there are two roots (A, B) to this quadratic equation. Differentiating equation (4.6) will provide the location of the point of inflexion, thus:

$$\partial_2 C/\partial P^2 = -2a_1 + 6a_2P = 0$$

from which

$$P = a_1/3a_2 \qquad (4.7)$$

The minimum heat rate occurs when, differentiating equation (4.4):

$$\partial H/\partial P = -a_1 + 2a_2P = 0$$

or

$$P = a_1/2a_2 \qquad (4.8)$$

This suggests that the point of inflexion will occur at a lower load than the minimum heat rate.

4.5 Optimal Trajectories

In some European power plants it is not unusual for the units within the plant to be optimally dispatched by the plant itself rather than the central grid dispatcher, who only assigns total plant load. When this occurs, it is possible not only to optimize the distribution of load among the plants as a function of load, but also to calculate the optimal trajectory each unit should follow during a plant load transition.

It is clearly possible to calculate, regress and plot the unit load versus the corresponding incremental cost and to do this individually for all the units in the plant. If IC is the incremental cost and MW the unit load, then the models that will be plotted have the form:

$$MW = a_0 + a_1 IC + a_2 IC^2 + a_3 IC^3 \qquad (4.9)$$

A simple search technique (Newton-Raphson, Regula Falsi, etc.) can then be used to find that incremental cost for all units at which the sum of the unit loads is equal to the total load required to be delivered by the plant.

The same procedure may also be used to find that incremental cost corresponding to the present load (P) plus X MW as well as the present load plus Y MW. There are now three total plant loads and three corresponding incremental cost values, from which the equivalent load from each unit can be computed using the corresponding equation (4.9) model shown above. These three data sets can now be used to compute the coefficients in a quadratic relationship expressing the optimal trajectory (optimal unit load vs. total plant load), thus: If

MW_P = total plant load P
MW_{P+X} = total plant load P+X
MW_{P+Y} = total plant load P+Y
$MW1_P$ = load on unit #1 at load P
$MW1_{P+X}$ = load on unit #1 at load P+X
$MW1_{P+Y}$ = load on unit #1 at load P+Y

Then, for unit #1,

$$MW1_P = c_0 + c_1 MW_P + c_2 MW_{p^2} \qquad (4.10)$$

$$MW1_{P-X} = c_0 + c_1 MW_{P+X} + c_2 MW_{P+X^2} \qquad (4.11)$$

$$MW1_{P-Y} = c_0 + c_1 MW_{P+Y} + c_2 MW_{P+Y^2} \qquad (4.12)$$

The values of constant c_0, c_1 and c_2 can now be calculated by Gaussian elimination and then later used to calculate the load on unit #1 as a function of the total plant load. The procedure would be repeated for units 2 thru N.

4.6   The Kalman Filter as a Process Modeling Tool

Some early vendors of industrial energy systems attempted to find the optimal assignments by actually perturbing the process and trying to estimate the direction of the next change(s) from the measured responses. This was time consuming, disturbed the operation, and was prone to error, especially from the false and often contradictory responses frequently experienced during transients. A prominent example of this occurs when the load on a boiler changes, the calculated efficiencies tending to be artificially high as the load falls, and vice versa, until a steady state has been established once more.

To avoid such problems, to provide a system that is self-learning and to allow the system to take up optimal assignments as quickly as possible after changes in demand have been detected, it is necessary to adopt an adaptive mathematical modeling approach. To keep the models in step with reality, the characteristic curves of all equipment should be periodically updated through the on-line regression analysis of data collected in a routine manner. Special filtering techniques must also be incorporated in these programs to reject the false data sets acquired during transients. The age of the data should also be taken into account.

One method of updating models much used in the past has been the *off-line* regression of carefully collected batches of data sets on which that model is to be based, the data sets being regressed off-line using a multiple linear regression program. These data sets must also be obtained with the equipment operating over a wide range of loads and operating conditions. Even so, the data always contains either instrument or process noise which adds an element of uncertainty.

This whole activity can thus become very tedious and time-consuming, while the hand-inspection of data sets and the rejection of those that distort the expected model is often the case. Clearly the use of some other technique requiring less human intervention and involvement would be very desirable.

The method outlined below is designed to meet these needs. It is based on the Recursive Least Squares (RLS) principle and uses the Kalman filter (Kalman (1964)) algorithm to model the equipment characteristic curve. The method accepts currently available data sets and there is no need to store the data permanently unless the data is required by some other task within the system. One advantage of the method is that it tends to perform better when the data is changing. Furthermore, having certain statistical properties, the algorithm is able to converge on a solution even in the presence of noise.

Models on which this technique is to be used are assumed to have a linear characteristic. However, should non-linearity be involved, techniques are available to transform the data so that the linear recursive least squares principle can be maintained.

While the Kalman filter can be very robust, there are times when the solution tends to diverge. This is especially liable to occur when the plant operates for too long a period under steady state conditions. Recognizing this tendency, the modeling procedure should incorporate various checks on the incoming data together with rules for rejecting certain kinds of data sets. It should also continuously apply artificial noise to all the independent variable signals.

As presented by Åstrom et al (1977), the Kalman filter was originally used to automatically adjust the tuning constants in sampled data process control loop algorithms. It was later adapted for use as an on-line modeling technique. Assume that it is desired to find the current characteristic curve model for a steam turbogenerator being of the form:

$$T = \theta_0 1 + \theta_1 EX + \theta_2 PWR \qquad (4.13)$$

where T is throttle flow, EX is extraction flow, PWR is power and θ_0, θ_1 and θ_2 are regression coefficients. The major data elements used in the Kalman filter algorithm are:

φ = input array of independent variables EX and PWR, including any preconditioning such as the addition of artificial noise. Note that $\varphi(1)$ contains the value of unity

θ = output array containing the set of presently calculated model parameter values (regression coefficients) for the linear model

K = Kalman gain array – used to modify the θ array in response to any error found in the measured value of T and its value predicted using the input data

R = Covariance matrix

P = Inverse of the last updated covariance matrix

T = predicted value of dependent variable calculated from EX and PWR

E = Error between measured T and predicted value of T, obtained by multiplying the last updated θ array by the new φ array

λ = Forgetting factor, tending to diminish the influence of older data

b_0 = Constant, set to unity in this application

Based on Åstrom et al (1977), the Kalman filter algorithm itself has the following form:

$$\theta_{(t)} = \theta_{(t-1)} + K_{(t-1)} * E_{(t-1)}/b0 \qquad (4.14)$$

$$K(t) = \frac{P_{(t-1)} * \varphi_{(t)}}{1 + (\varphi_{(t)}T * P_{(t-1)} * \varphi_{(t)} - 1)/b_0} \qquad (4.15)$$

$$R_{(t)} = \lambda * R_{(t-1)} + \varphi_{(t)}{}^{T} * \varphi_{(t)} \qquad (4.16)$$

$$P_{(t)} = R_{(t)}{}^{-1} \qquad (4.17)$$

$$E_{(t)} = T_{(t)} - \theta_{(t)}{}^{T} * \varphi_{(L)} \qquad (4.18)$$

4.7 Conclusions

The task of optimizing the distribution of energy within an industrial plant can take many forms. There are also a variety of optimization techniques available for their solution. The selection of the most appropriate technique depends on the equipment models included in the network constructed during the analysis, especially whether they are linear, non-linear or a mix of the two. In some plants the total problem may even be divided into several parts, a different technique being used to solve each sub-problem.

Chapter 5
Applications and Case Studies

A number of strategies were identified in chapter 4 that have been successfully used to optimize the steam and power distributions within the energy systems of several industrial plants. Chapter 5 will be devoted to reviewing how these techniques were applied to a number of actual cases. In almost every case, existing plant data was studied to estimate the magnitude of the savings that might be realized from the application of an optimization strategy. In the course of these studies, there was usually something unique to be learned about the intrinsic behavior of the plant or system that may have a more general application. Attention will be drawn to these insights where appropriate.

5.1   Linear Programming Applications

Figure 4.1 shows how the energy-optimizing task for an industrial plant can be decomposed into a set of sub-optimizing problems, each with its own characteristics. Thus, while the optimizing of live-steam generation is a non-linear problem, experience has shown that the optimizing of the cogeneration of power and process steam is essentially linear, allowing linear programming techniques to be used for their solution.

5.1.1   Plant with Only One Steam Turbogenerator

Figure 4.2 shows the flow sheet for a plant with only one steam turbogenerator while the corresponding L.P. matrix is

defined in Table 4.1. The system was designed not only to satisfy the total amount of energy consumed by the plant; but
also to minimize the combined cost of purchased boiler fuels
and utility power by optimizing the distribution of energy
resources within the network. Although this was a very simple
system, it was only during the study that the significance of
the desuperheating water injected after pressure reducing valve
PRV1 was realized and changed the way in which the power
house was operated. Originally, the high pressure boiler was
only used to supply steam to the turbogenerator while the
low pressure boiler was used to supply process steam. It was
thought that if the high pressure steam were used to supply
steam to the process via PRV1, its higher steam enthalpy
would make it more expensive than the low pressure steam
allowing it to be used only in emergencies. In any case, boiler
feedwater had been the source of water for the desuperheater
and it had to be taken out of service frequently due to fouling
of the ports from deposition of dissolved solids. Eventually it
was shut off completely.

However, the study showed that with the injection of desuperheating water at a rate of 13% of the live steam flow,
the increased mass flow leaving PRV1 now had a unit cost
equal to, or even less than, that of the low pressure steam.
With this new insight, condensate drawn from the condenser
now became the source of desuperheating water and the
desupeheater was brought back into service. The tendency
now is to reduce the load on the less efficient low pressure
boiler and utilize as much as possible of the balance of the
capacity of the high pressure boiler. In addition to the savings
from optimizing the distribution of the energy resources
within the network, this change in operating strategy made
its own significant contribution to the total savings that
were realized from the installation of the energy optimizing
system.

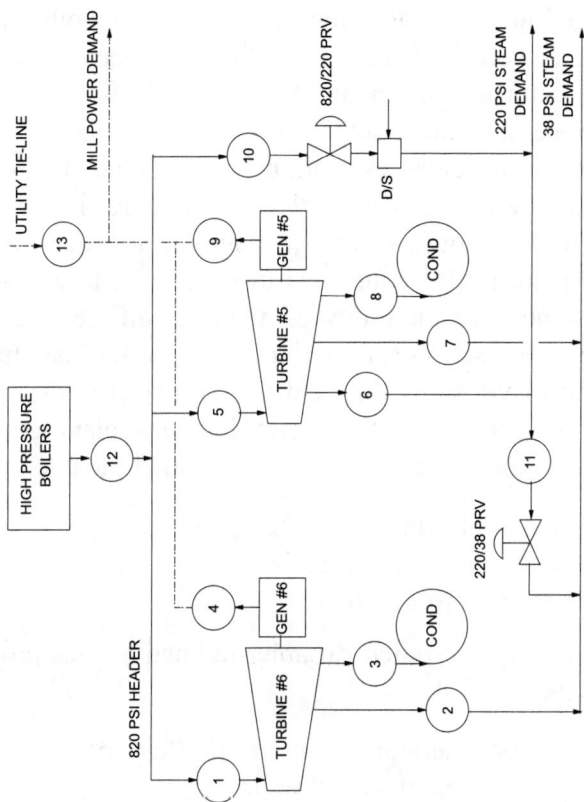

Figure 5.1   Cogeneration flowsheet for KVP plant

5.1.2 Paper Mill with Two Steam Turbogenerators

The flowsheet for this plant, originally described by Smith and Putman (1988), is shown in Figure 5.1 while the L.P. matrix is defined in Table 5.1. The plant was provided with two condensing/extraction steam turbogenerators, each equipped with a load governor as well as with extraction governors, the latter arranged generally as shown in Figure 3.7. Both turbo-generators received live steam from the boiler header, while Turbine #6 extracted steam at 38 psi and Turbine #5 was provided with extraction at both 220 psi as well as at 38 psi. In the event of insufficient extraction steam being available, pressure reducing valves ensured that the pressures in the process steam headers would always be maintained at their design values. The plant is also connected to the local utility tie-line and it is assumed that the utility contract has different energy rates for on-peak vs. off-peak periods and that the maximum demand limit may also be different for the two periods.

The current set of the three simultaneous plant energy demand targets that always have to be satisfied include:

- Process steam consumption at 220 psi
- Process steam consumption at 38 psi
- Electric power consumption

The set of energy resource variables defined for this project is as follows:

1. THR6 - Turbogenerator #6: - Throttle flow lb/h
2. LPX6 - - 38 psi extraction flow lb/h
3. COND6 - - Condenser flow lb/h
4. PWR6 - - Generated power Kw
5. THR5 - Turbognerator #5: - Throttle flow lb/h
6. HPX5 - - 220 psi extraction flow lb/h
7. LPX5 - - 38 psi extraction flow lb/h
8. COND5 - - Condenser flow lb/h
9. PWR5 - - Generated power lb/h

10. PRV8 - Steam flow to 820/220 pressure reducing valve lb/h
11. PRV2 - Steam flow to 220/38 pressure reducing valve lb/h
12. BLRS - Total live steam flow from boilers lb/h
13. TIE - Power drawn from utility tie-line Kw

One of the variables contained in the solution for a given set of energy demands is the total live steam flow from the high pressure boilers: this can become the target in a separate optimizing task for optimizing the fuel and steam flow distributions for these boilers. Meanwhile, the set of equality and constraint equations defined for this project is as follows:

1. Turbogenerator #6: Throttle flow equation
2. Turbine steam balance
3. Maximum throttle flow rate
4. Maximum 38 psi extraction flow rate
5. Maximum generated power
6. Minimum generated power
7. Minimum condenser flow rate
8. Turbogenerator #5: Throttle flow equation
9. Turbine steam balance
10. Maximum throttle flow rate
11. Maximum 38 psi extraction flow rate
12. Maximum generated power
13. Minimum generated power
14. Minimum condenser flow rate
15. Maximum tie-line power
16. Minimum tie-line power
17. 820 psi steam header balance

Equations for plant energy demand targets (net targets stored in RHS column of these equations):

18. Total power consumed by plant
19. Amount of 38 psi process steam consumed by plant
20. Amount of 220 psi process steam consumed by plant

Table 5.1 LP matrix-KVP plant

MIN	1 THR6	2 LPX6	3 COND6	4 PWR6	5 THR5	6 HPX5	7 LPX5	8 COND5	9 PWR5	10 PRV8	11 PRV2	12 BLRS	13 TIE		RHS
COST												.0067	.0435		
1	1	-.5												=	12400
2	1	-1	-1											=	0
3	1			-8.37										≤	135000
4					1	-.75	-.5		-8					≤	130000
5					1	-1	-1							≤	9500
6			1											≥	3000
7		1					1							≥	10000
8														=	12000
9														=	0
10						-1	-1			1	1			≤	240000
11										1	1			≤	150000

							Rel.	RHS
12	1						$\leq$	9500
13						1	$\geq$	2000
14			1				$\geq$	10000
15	1	1					$\leq$	9000
16				1		1	$\geq$	2000
17	1				1		$=$	0
18		1		1.124		1	$=$	18519
				-1	-1		$=$	126850
						-1	$=$	169820

Costs are applied only to the steam generated by the boilers ($0.0067/lb) and tie-line power (0.0435/kwh). It is assumed that the desuperheating water ratio for PRV820/220 is 12.4%. The matrix embodies several practical considerations. For instance, the steam flow to a condenser should not be allowed to fall below a minimum value if the backpressure is to remain stable. Further, the power drawn from the tie-line should not be allowed to fall below a minimum value in order to reduce the risk of an unexpected trip of the tie-line breaker. The sources of the total amount of power consumed by the plant are:

• Minimum condensing power—power generated as a function of the minimum steam flow to the condenser
• Forced condensing power
• Power cogenerated by extraction steam
• Power purchased from the tie-line.

So long as a turbogenerator is operating, a certain amount of power has to be generated as a function of the minimum steam flow to the condenser and this must be consumed by the plant, regardless of cost. Any power generated from condenser steam in excess of this minimum value (forced condensing power) is to some extent optional and whether it is generated often depends on its cost with respect to the cost of tie-line power. Meanwhile, the power cogenerated with the extraction of steam is relatively inexpensive and is to be encouraged.

Figure 5.2 contains plots of these different types of power throughout a typical day, if tie-line power is not being regulated. Mill deficit power is defined as the difference between total plant demand and the sum of the power generated from the minimum condensing steam flow and extraction steam flows. This deficit may be satisfied by a judicious combination of forced condensing and purchased power, the amounts based on considerations of least total cost. Figure 5.3 shows a similar plot in which the amount of forced condensing power has been

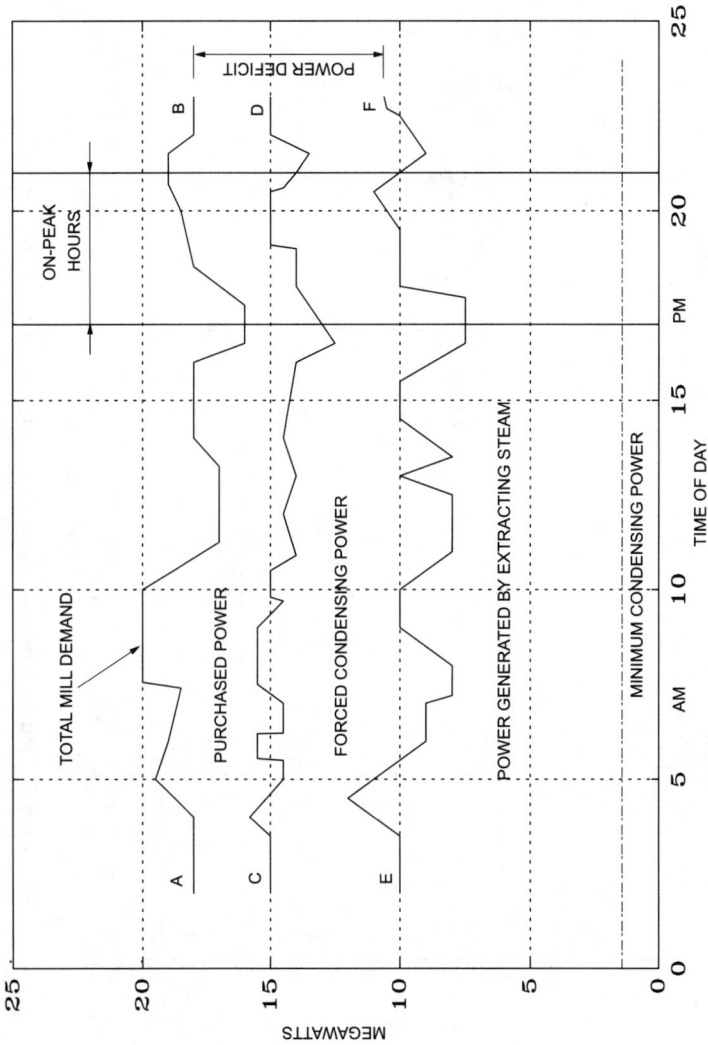

Figure 5.2   Plot of power consumption - no tie-line control

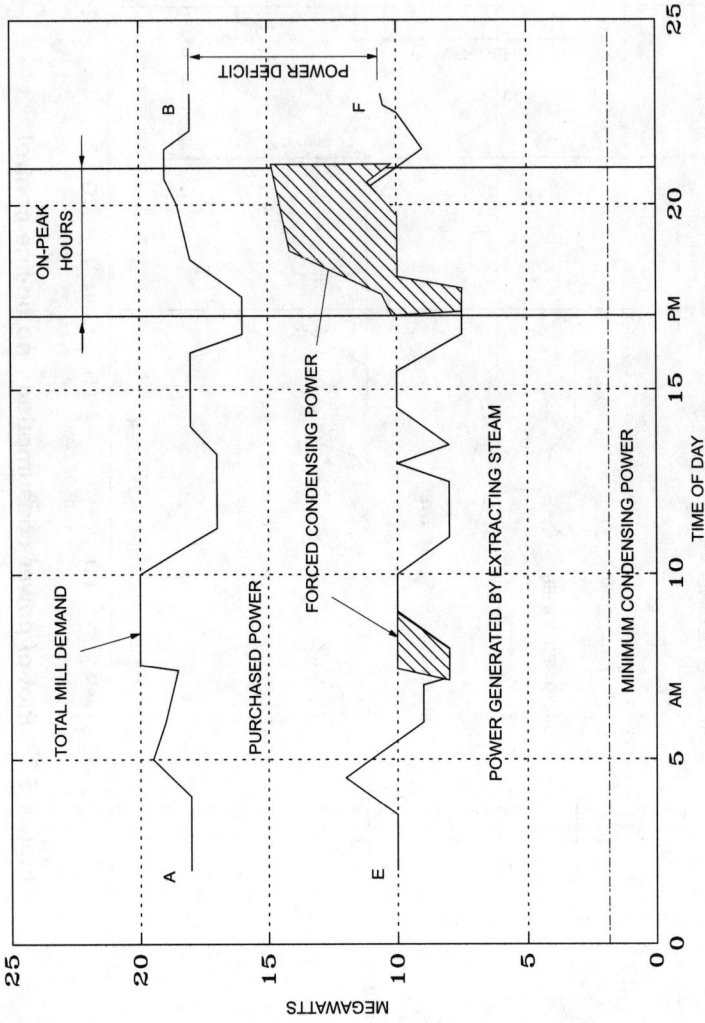

Figure 5.3   Plot of power consumption - with tie-line control

minimized, the difference being made up by cheaper purchased power, the maximum tie-line power being regulated by adjusting the amount of forced condensing power to maintain the tie-line limit established during previous on-peak demand periods.

Before exercising the L.P. matrix to calculate the optimum energy distribution for the current set of demand targets, the cost of steam and power should be updated to reflect current costs with particular regard to whether the plant is operating in an on-peak or off-peak period. The maximum tie-line power should also be updated to reflect the type of period while the RHS targets for equations 18 thru 20 should be updated to reflect current values.

The results from running the matrix shown in Table 5.1 are given in Table 5.2 below, one with a tie-line power demand limit of 9000 kw and the other with a limit of 4000 kw.

Given the power and process steam demand targets and a tie-line limit of 9000 kw, the targets are satisfied at least cost by operating both turbogenerators under minimum condensing power but with Turbogenerator #5 also operating at maximum throttle flow. The power being drawn from the tie-line is only 6004 kw compared with the limit of 9000 kw. The demand for 220 psi steam is being satisfied entirely by extraction from TG#5 and, as should be expected, under these conditions no steam is being drawn through either pressure reducing valve.

If the tie-line limit is now reduced to 4000 kw to reflect an on-peak period, the power from both turbogenerators is increased to offset the reduced draw from the tie-line. However, while Turbogenerator #6 is still operating under minimum condenser conditions, the condenser flow in Turbogenerator #5 has been increased until it is now generating its maximum load of 9500 kw. The distribution of 38 psi extraction flows has also changed.

Smith and Putman (1988) calculated the annual energy savings realized from minimizing forced condensing power to be $751,056 from a plant operating for 300 days per year.

Table 5.2   Results for two extraction/condensing turbogenerators

Variable	9000 kw	4000 kw	Constraint
THR6	76670	98818	
LPX6	66670	88818	
COND6	10000	10000	≥ 10000
PWR6	3695.9	5019	
THR5	240000	234380	≤ 240000
HPX5	169820	169820	
LPX5	60180	38031	
COND5	10000	26529	≥ 10000
PWR5	8818.2	9500	≤ 9500
PRV820/220	0.	0.	
PRV220/38	0.	0.	
BLRS	316670	333199	
TIE	6004.94	4000	≤ 4000
Power Target	18519	18519	
38 psi Steam Target	126850	126850	
220 psi Steam Target	169820	169820	
Cost	$2382.9/h	$2406.43/h	

One of the peripheral savings resulted from being able to reduce the number of condenser circulating water pumps that needed to be operated for most of the year. However, care had to be exercised to avoid silting of the condenser tubes when running with lower tube water velocities.

5.1.2.1   An Off-Line Study

A small paper mill in Georgia operated as an island independent of the local utility. Data from the power house was analyzed and it was found that the characteristic curves for the two turbogenerators had changed, in one case due to wear of the high pressure steam nozzles and, in the other, to some damage to the low pressure blading. After generating new characteristic curves, developing the L.P. matrix for this plant and calculating the cost for each of several sets of operating conditions, the linear program was run for each of the set of target demands and the results analyzed. The subsequent report made clear that the assignments of steam and power between the turbogenerators could be improved and substantial savings anticipated.

Some weeks later, new sets of current operating data were presented to the program but the magnitude of the savings had decreased significantly. An investigation showed that the plant had applied the recommendations contained in the earlier report and had realized almost all the savings by following manually the trends identified in that report.

Today, the L.P. matrix could be exercised on a personal computer located in the control room and the recommendations implemented manually by studying the results.

5.1.3 Plant with One Condensing and One Backpressure Turbogenerator

Figure 5.4 shows a different plant with only one turbogenerator provided with a condenser, the other being a back pressure machine. With the former, the extraction flow and power are not intimately related because the exhaust flow to the condenser can be adjusted to generate more or less power. However, with a back pressure machine, the amount of power generated is directly coupled to the magnitudes of the extraction and exhaust flow rates, the values needed to maintain the process steam header pressures in response to the process steam demands.

However, it will be shown that, by deliberately increasing the exhaust flow rate, in this case by venting low pressure steam, the power generated by the back pressure machine can be increased. While this might seem to be uneconomic, it should be remembered that in a condenser only the sensible heat is recovered from the exhaust steam, the latent heat being

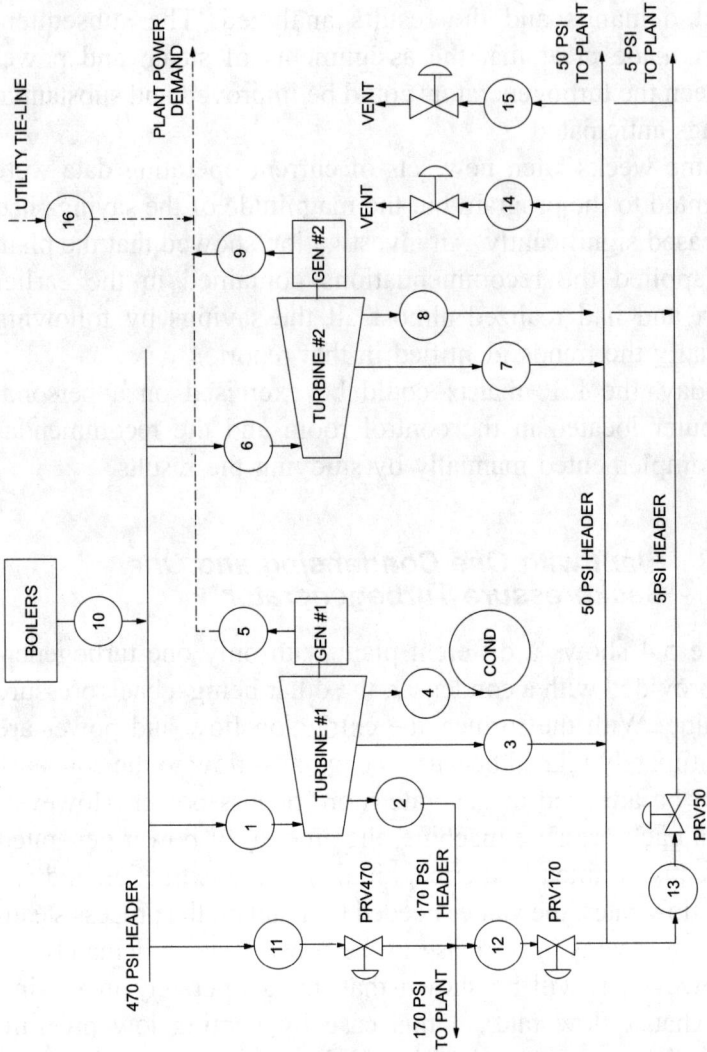

Figure 5.4 Cogeneration flowsheet for Penna plant

discharged into the atmosphere. Thus, while generating power by venting steam may cost more than the power generated by condensing, the increased cost may be justified when the cost of utility power becomes high, e.g. during on-peak periods. As a matter of interest, in a large chemical plant in Rochester, NY it is common practice to generate power in this way during a significant part of the day.

The set of energy resource variables defined for the project shown in Figure 5.4 was as follows:

1. THR1 - Turbogenerator #1: - Throttle flow lb/h
2. HPX1 - 170 psi extraction flow lb/h
3. LPX1 - 50 psi extraction flow lb/h
4. COND1 - Condenser flow lb/h
5. PWR1 - Generated power Kw
6. THR2 - Turbogenerator #2: - Throttle flow lb/h
7. IPX2 - 50 psi extraction flow lb/h
8. LPX2 - 5 psi exhaust flow lb/h
9. PWR2 - Generated power lb/h
10. BLRS - Steam flow from boilers lb/h
11. PRV470 - Steam flow to 470/170 pressure reducing valve lb/h
12. PRV170 - Steam flow to 170/50 pressure reducing valve lb/h
13. PRV50 - Steam flow to 50/5 pressure reducing valve lb/h
14. VENT50 - Steam flow to 50 psi vent valve lb/h
15. VENT5 - Steam flow to 5 psi vent valve lb/h
16. TIE - Power drawn from utility tie-line Kw

Meanwhile the set of equality and constraint equations defined for this project was as given below:

Y.1   Characteristic equation for TG#1: THR1 − 0.8 HPX1 − 0.625 LPX1 − 12.4 PWR1 = 7081
Y.2   Steam balance for TG#1: THR1 − HPX1 − LPX1 − COND1 = 0
Y.3   Maximum throttle flow THR1 ≤ 200000

Y.4   Maximum condenser exhaust flow COND1 $\leq$ 26000

Y.5   Minimum condenser exhaust flow COND1 $\geq$ 4000

Y.6   Maximum generated power PWR1 $\leq$ 5000

Y.7   Minimum generated power PWR1 $\geq$ 1000

Y.8   Characteristic equation for TG#2 THR2 $-$ 0.35 IPX2 $-$ 15.7 PWR2 = 7081

Y.9   Steam balance for TG#2 THR2 $-$ IPX2 $-$ LPX2 = 0

Y.10 Minimum generated power PWR2 $\geq$ 1000

Y.11 Minimum exhaust flow TG#2 LPX2 $\geq$ 14000

Y.12 Maximum throttle flow THR2 $\leq$ 180000

Y.13 Maximum generated power PWR2 $\leq$ 6000

Y.14 Maximum steam flow from boilers BLRS $\leq$ 330000

Y.15 Maximum tie-line power TIE $\leq$ 5000

Y.16 470 psi header balance THR1 + THR2 $-$ 0.96 BLRS + PRV470 = 0

Y.18 Minimum boiler load $-$ BLRS BLRS $\geq$ 80000

Y.19 Minimum tie-line power TIE $\geq$ 1000

Y.20 Power demand target PWR1 + PWR2 + TIE = 10471

Y.21 170 psi process steam target HPX1 + PRV470 $-$ PRV170 = 26500

Y.22 50 psi process steam target LPX1 + IPX2 $-$ 0.012 BLRS + PRV170 $-$ PRV50 $-$ VENT50 = 101500

Y.23 5 psi process steam target LPX2 + 0.0385 BLRS + PRV50 $-$ VENT5 = 46300

5.1.3.1 Costs

BLRS = $0.0034/lb steam

VENT50 = $0.0027/lb steam

VENT5 = $0.0026/lb steam

TIE = $0.06/kwh

5.1.3.2 Comments

This matrix shows that it is not necessary to state all the parameters in the system specifically. For instance, in equation

Y.16 there is a coefficient of 0.96 before the BLRS term, indicating that 4% of the steam from the boilers is supplied to the feedwater heaters. Similarly a coefficient of 0.0385 has been applied to the term BLRS to allow for the amount of 5 psi steam that flashes off from the deaerators. The results from exercising this matrix are given in Table 5.3 below.

Column #1 shows the results if convergence is allowed to occur naturally. It is seen that the steam flows and power assigned to turbogenerator #1 to meet the set of demand targets is constrained by the condenser flow rate reaching its maximum value. Column #2 shows the effect of forcing the power generated by the backpressure machine to its maximum value. As a result, the tie-line power decreases by 454.7 kw while the total cost rises from $868.69 to $885.96, an increase of $17.27. This means that the incremental power costs only $0.038/kw; which is much less than the cost of tie-line power, given as being $.06/kwh. In this plant the price of purchased power would have to drop below $0.038/kwh before venting would be aborted.

Thus, even though the steam is vented to atmosphere and there is some cost of replacing it as makeup water well as the heat lost, it is still cheaper than buying power from the utility, the only other option available.

There may, of course, be other considerations when determining how a plant should best be operated. In this case, with both turbogenerators pushing upper constraints in trying to meet the desired combination of process steam flow and power demands, the load on the tie-line can only be controlled by switching off electrical loads. This may not be desirable so that the turbogenerators may have to be operated below their upper constraints in order to leave some cushion for tie-line control. Thus the mathematical optimum may have to be over-ridden in order to provide some flexibility to the operators.

From the above it should be clear that much useful information can be obtained by constructing the linear programming matrix for a given plant and exercising it in various ways.

Table 5.3   Effect of forced venting of low pressure steam

Parameter Column	Natural Convergence 1	PWR2 forced to 6000 kw 2	Constraint 3
THR1	52500	52500	
HPX1	26500	26500	
LPX1	0.	0.	
COND1	26000	26000	≤ 26000
PWR1	1953.1	1953.1	
THR2	142419	149589	
IPX2	103936	104026	
LPX2	38482.9	45563	
PWR2	5545.33	6000	≤ 6000
BLRS	203041	210509	
PRV470	0	0	
PRV170	0	0	
PRV50	0	0	
VENT50	0	0	
VENT5	0	7367	
TIE	2972.5	2517.8	
Power target	10471	10471	
170 psi steam target	26500	26500	
50 psi steam target	101500	101500	
5 psi steam target	46300	46300	
Cost	868.69	885.96	

5.1.4   Optimal Utilization of Waste Heat Steam—Phosphate Processing Plant

To demonstrate the generality of linear programming to the optimization of power house optimization problems, Putman and Panizza (1984) discussed its application to the optimization of waste steam utilization within a phosphate processing plant located in Florida, the simplified flow sheet for which is shown in Figure 5.5. In this plant, a large amount of high pressure saturated steam was being generated in a part of the main process. Much of this steam was then superheated, vent valve (VNT575) being provided to maintain a positive differential pressure across the superheater. In addition, the power house was equipped with one back pressure turbogenerator (TG#1) and one extraction/condensing steam turbogenerator (TG#2), both operating in parallel with the local tie-line. Thus, so long as sufficient steam is being drawn by TG#1 and/or TG#2, VNT575 should be closed. There were also a number of lower process steam headers each with its own independent steam demand, all of which had to be satisfied.

To maintain the pressure in the 600 psi header and provide a source of 250 psi steam to the process, a portion of the waste heat steam was let down through valve(s) LDV600. In the event of the demand for 250 psi steam being less than the surplus let down through valve(s) LDV600, another let down valve (LDV250) was provided so that this steam could be spilled over into the 45 psi system. Meanwhile, the 45 psi system was provided with a fin-fan condenser intended to recover as much of any surplus steam as possible in the form of condensate, any balance being vented to atmosphere through valve VENT45.

It should also be noted that valves LDV600 and LDV250 are primarily let down valves controlled to maintain the pressure upstream of these valves at some desired value. However, when necessary, they could also function as pressure reducing valves, the controls being switched to maintain the pressure downstream of these valves at some desired value.

Clearly, the primary optimization task is to satisfy the combined demands for process steam and power at least cost. In this case it involves passing steam to the turbogenerators to minimize tie-line power costs while also making sure that the amount of 45 psi steam recovered is maximized and the

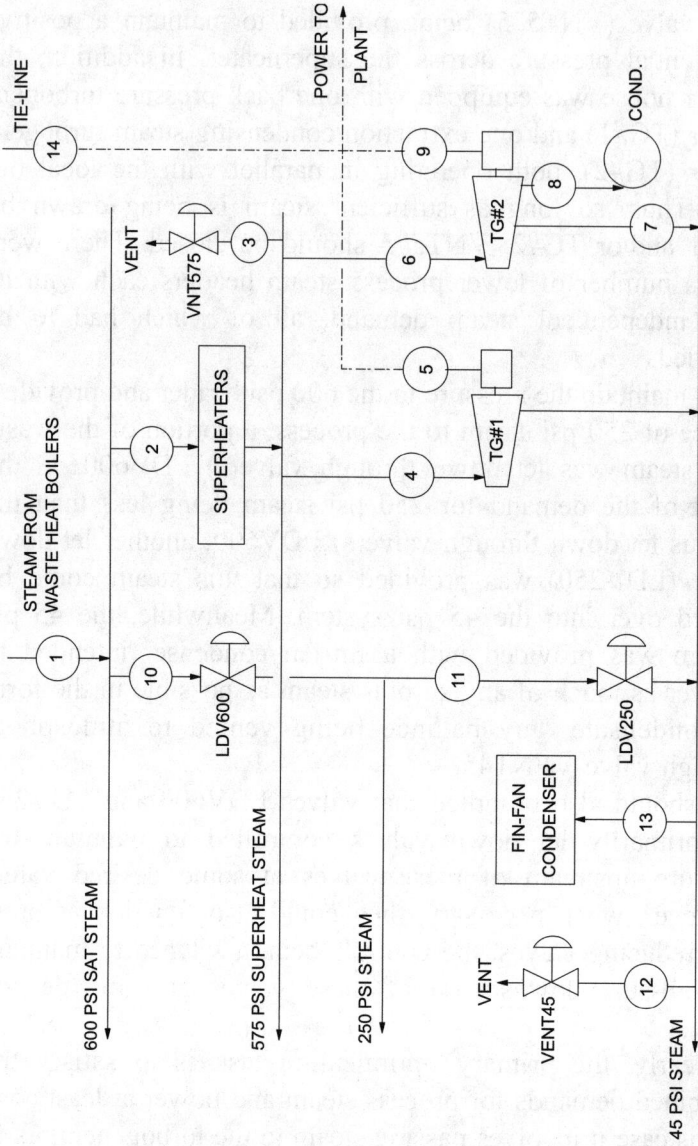

Figure 5.5   Steam distribution flow sheet - Phosphate plant

amount of steam vented to the atmosphere is minimized. The parameter set defined for this project was as follows:

5.1.4.1   Variables

1. WHBLRS 660 psi steam from waste heat boilers lb/h
2. SUPHTD 575 psi steam from superheaters lb/h
3. VNT575 575 psi Vent Steam lb/h
4. THR1 Turbogenerator #1: Throttle flow lb/h
5. PWR1 Power kw
6. THR2 Turbogenerator #2: Throttle flow lb/h
7. LPEX2 45 psi extraction flow lb/h
8. COND2 Condenser flow lb/h
9. PWR2 Power kw
10. LDV600 600/250 psi letdown valve flow lb/h
11. LDV250 250/45 psi letdown valve flow lb/h
12. VENT45 45 psi vent valve flow lb/h
13. FFCOND Flow to fin-fan condenser lb/h
14. TIE Tie-line power kw

Meanwhile, the set of equality and constraint equations defined for this project was as given below:

Y.1   Upper constraint on 575 psi vent valve VNT575 $\leq$ 310,000
Y.2   Maximum power from TG#1 PWR1 $\leq$ 10,000
Y.3   Minimum power from TG#1 PWR1 $\geq$ 1,000
Y.4   Characteristic equation – TG#1 THR1 – 20 PWR1 – 34,000
Y.5   Maximum throttle flow – TG#2 THR2 $\leq$ 650,000
Y.6   Maximum extraction flow – TG#2 LPX2 $\leq$ 430,000
Y.7   Maximum condenser flow – TG#2 COND2 $\leq$ 430,000
Y.8   Maximum power from TG#2 PWR2 $\leq$ 39,300
Y.9   Minimum power from TG#2 PWR2 $\geq$ 5,000

Y.10 Characteristic equation – TG#2 THR2 – 0.5435 LPEX2 – 8.654 PWR2 = 61475

Y.11 Steam balance – TG#2 THR2 – LPEX2 – COND2 = 0

Y.12 Maximum flow – LDV600 LDV600 ≤ 500,000

Y.13 Maximum flow – LDV250 LDV250 ≤ 500,000

Y.14 Maximum flow – VENT45 VENT45 ≤ 100,000

Y.15 Maximum flow – FinFan Condenser FFCOND ≤ 100,000

Y.16 Maximum tie-line power TIE ≤ 12,000

Y.17 Minimum tie-line power TIE ≤ 2,000

Y.18 600 psi sat. steam target WHBLRS – SUPHTD – LDV600 = 71,700

Y.19 575 psi sup. steam target SUPHTD – VNT575 – THR1 – THR2 = 210,000

Y.20 250 psi steam target LDV600 – LDV250 = 22,200

Y.21 45 psi steam target THR1 + LPEX2 + LDV250 – VNET45 – FFCOND = 184,000

Y.22 Plant power demand target PWR1 + PWR2 + TIE = 28,403

Y.23 Waste heat steam available WHBLRS = 776,700

5.1.4.2   Costs

VNT575 $0.0065/lb
COND2 $0.004/lb
VENT45 $0.006/lb
FFCOND $0.004/lb
TIE $0.0595/kwh

5.1.4.3   Optimization

The optimal values of the variables corresponding to a given set of process steam and power demands and the associated amount of available waste heat steam is given in Table 5.4.

Table 5.4.   Optimal steam distribution

Parameter	Optimal Results	Constraints
WHBLRS	776700	= 776,700
SUPHTD	647246	
VNT575	0	
THR1	234000	
PWR1	10000	≤ 10,000
THR2	203426	
LPEX2	0	
COND2	203426	
PWR2	16403	
LDV600	57573	
LDV250	35373	
VENT45	0	
FFCOND	85373	
TIE	2000	≥ 2000
600 psi steam target	71700	
575 psi steam target	210000	
250 psi steam target	22200	
45 psi steam target	184000	
Power target	28403	
Cost	1274.2/h	

5.1.4.4   Comments

Every linear programming matrix used for optimizing industrial energy systems is site specific, not only with regard to the set of equalities and constraints involved but also the set of variables selected for the cost function, as well as the actual costs assigned to each. In this case, purchased power is clearly a cost to be minimized and the cost assigned is that contained in the utility contract for that hour and day. Condensers reject heat and this rejection should be minimized; while the cost of venting steam includes not only the loss of its heat but also the replacement cost of the water. Thus the cost of venting steam is greater than that of passing steam to a condenser.

In the on-line control of power house optimization, the current set of steam and power demands is first calculated and stored in the appropriate elements in the matrix, together with the amount of waste heat steam available. The L.P. is then exercised and the results analyzed. For this combination of demands and available steam the optimal distribution is to minimize the amount of power purchased from the tie-line by generating internally as much power as possible. Thus, the tie-line operates at its minimum draw while TG#1 is being run at its maximum power, the balance of the power being supplied by TG#2. However, this is all condensing power since, with TG#1 at full load, there is more than enough 45 psi steam to satisfy the process demand at that pressure. In fact, there is a surplus of 45 psi steam that is passed to the fin-fan condenser for recovery as condensate.

Furthermore, the amount of steam drawn through the super-heaters is not sufficient to avoid some of the 600 psi waste heat steam having to be let down to lower pressures and in a quantity greater than the demand for 250 psi steam. This means that LDV250 is also passing surplus steam into the 45 psi header which, combined with the 45 psi exhaust from TG#1, means that no 45 psi needs to be extracted from TG#2.

A potential problem with L.P. solutions is that, often, several variables are forced to their upper or lower limits in the solution. While this is the true mathematical optimum, the operators may prefer, for instance, a turbogenerator not be run flat out but be set to a lower load so as to provide some cushion for the plant in the event of a tie-line trip. For this reason, all of the plants that have used this technique have collected the appropriate set of steam and power demands for a wide variety of historical cases and studied, off-line, the consequences of optimizing each. From this analysis, not only can the reduction in operating costs be estimated but also rules generated as to how the plant should be operated over a broad spectrum of historical operating conditions, not excluding the possibility of modifying or constraining some of the variables in the optimal solution.

5.2   SSDEVOP Applications

The Simplex Self-Directing Evolutionary Operation (SSDE-VOP) technique was introduced in Section 4.2. It is able to solve both linear and non-linear optimizing problems and an example of a linear problem was included in that Section. It works well when the number of variables included in the experimental design is small and is then able to converge rapidly on the solution.

5.2.1   SSDEVOP Solution of Linear Problem

Section 4.2 reviewed a case in which the extraction flows and generated power had to be assigned between two turbogenerators with essentially linear characteristics, the flow sheet being shown in Figure 4.6. Although other variables were also involved, the experimental design itself, shown in Figure 4.5, included only three variables:

1. L.P. Extraction flow from turbogenerator TG2 - E2
2. Turbogenerator TG1 Condenser flow - C1
3. Turbogenerator TG2 Condenser flow - C2

The energy resource targets for this case included:

   a. Total power demanded by plant, including tie-line power - PWR

b. Total amount of L.P. process steam - LPTOT
c. Total amount of H.P. process steam - HPTOT

It was assumed that the high pressure process steam flow (HPTOT) would always be satisfied either by extracting this steam from tubogenerator #1 or by passing the required amount of live steam through pressure reducing valve PR1. A number of other variables may now be derived, reference being made to Equation (4.1), this being a typical characteristic equation for an industrial steam turbogenerator:

L.P. extraction flow from TG#1 $-$ E1 = LPTOT $-$ E2

$$(5.1)$$

Throttle flow to TG#! $-$ THR1 = HPTOT + E1 + C1

$$(5.2)$$

Throttle flow to TG#2 $-$ THR2 = E2 + C2   (5.3)

Power from TG#1 $-$ PWR1 = (THR1 $-$ K1$_1$

$$-K2_1 \; HPTOT - K3_1 E1)/K4_1 \quad (5.4)$$

Power from TG#2 $-$ PWR2 = (THR2 $-$ K1$_2$

$$-K3_2 E2)/K4_2 \quad\quad\quad (5.5)$$

Tie $-$ line power $-$ TIE = PWR$-$

PWR1 $-$ PWR2   (5.6)

Total live steam $-$ TOTSTM = THR1 + THR2

$$(5.7)$$

Cost $-$ COST = TIE*($/kwh) + TOTSTM*($/lb)

$$(5.8)$$

Before the optimizing procedure can begin, the following have to be defined:

i. Upper and lower constraints on each experimental design variable E2, C1 and C2

ii. Upper and lower constraints on THR1, THR2, PWR1, PWR2, TIE

iii. Initial perturbation to be applied to each design variable - lb/h. In this case, the initial perturbation (DEL) can be the same for all variables, so that

$$DEL = a1 = a2 = a3 = 1000 lb/h$$

iv. Feasible starting case. This can be the set of present values of variables E2, C1 and C2. Alternatively, E2 and C2 can be assigned values within their ranges, the value of TIE being set to TIE_{min}. Then:

$$E1 = LPTOT - E2 \tag{5.9}$$

$$PWR2 = ((E2 + C2) - K1_2 - K3_2 E2)/K4_2 \tag{5.10}$$

$$PWR1 = PWR - TIE_{min} - PWR2 \tag{5.11}$$

$$THR1 = K1_1$$

$$+ K2_1 HPTOT + K3_1 E1 + K4_1 PWR1 \tag{5.12}$$

$$C1 = THR1 - HPTOT - E1 \tag{5.13}$$

After completing these simple preparatory calculations, which determine the starting feasible set of values for $E2_{BASE}$, $C1_{BASE}$ and $C2_{BASE}$ (See Figure 4.5), these variables may then be perturbed by the selected value of DEL and the perturbed values stored in rows 1 thru 4 of the upper array. The cost of each row (B1, B2, B3 and B4) may now be calculated using Equations (5.1–5.8) above and the values stored in the right hand column of the upper array.

These four cases are now examined and the worst (highest cost) case identified. Assume this was Case #2. The SSDE-VOP procedure now requires that new base values of E2, C1 and C2 be calculated by:

- Calculating new mean values from Cases #1, #3 and #4
- Doubling these mean values
- Subtracting the worst case values from the doubled mean values

The latter ensures that the solution moves away from the worst cases. Penalties are also applied to the cost of a Case if any significant variable within that Case encroaches on its limits.

Convergence is defined as having occurred when the difference in base cost (i.e. the cost of Case #4 in the matrix) between two iterations of the optimizing procedure is less than an assigned amount. Should it be desired to obtain a more accurate convergence, the perturbation interval (DEL) may now be reduced. Using the previous set of converged (optimum) values as the new starting set, the procedure is repeated until a new convergence is obtained.

5.2.2   SSDEVOP Solution of Non-Linear (Boiler) Problem

It will have been noted that, with the linear programming problems described in Sections 5.1.1 and 5.1.2, only the total amount of live steam (BLRS) required to satisfy the solution was calculated, no attempt being made to determine the optimum assignment of loads among the boilers. Further, if the boilers were being fired with more than one fuel, no attempt was made to calculate the optimal assignments of each fuel to each boiler. The reason is that these are non-linear problems for the solution of which the SSDEVOP technique is uniquely suitable.

Assume that BLRS is to be generated in two boilers each able to be fired simultaneously with both fuel oil and natural gas. The structure of the SSDEVOP experimental design for this problem is shown in Figure 5.6 and involves only three significant variables, namely:

1. Load on boiler #1
2. Oil flow to boiler #1
3. Oil flow to boiler #2

All other significant variables can be derived from these three. To set up the starting feasible case requires the following steps:

a. Knowing the total amount of steam (BLRS) to be generated, the initial base value of BLR1 will be 50% of BLRS thus:

$$BLR1 = 0.5^*BLRS \qquad (5.14)$$

$$BLR2 = BLRS - BLR1 \qquad (5.15)$$

b. The combustion efficiency vs. load model for each boiler and each fuel will have been already determined by the regression analysis of data taken during site tests. Assume that the initial base oil flows will be that equivalent to 50% of each boiler load. If HTACQU is the heat acquired per lb steam, heating value of gas is HVG and that of oil HVO, then:

$$OIL1 = 0.5^*BLR1^*HTACQU/$$

$$(0.95^*HVO^*f(EFFO1(BLR1))) \qquad (5.16)$$

VARS.	BLR1	OIL1	OIL2	COST
1	$BRL1_{BASE} - a1$	$OIL1_{BASE} - a2$	$OIL2_{BASE} - a3$	B1
2	$BRL1_{BASE} + a1$	$OIL1_{BASE} - a2$	$OIL2_{BASE} - a3$	B2
3	$BRL1_{BASE}$	$OIL1_{BASE} + 2*a2$	$OIL2_{BASE} - a3$	B3
4	$BRL1_{BASE}$	$OIL1_{BASE}$	$OIL2_{BASE} + 3*a3$	B4

5	AVERAGE VALUES CASES 1, 3 and 4
6	TWICE AVERAGE VALUES CALCULATED IN STEP 5
7	SUBTRACT VALUES OF WORST CASE #2

NEW BASE CASE

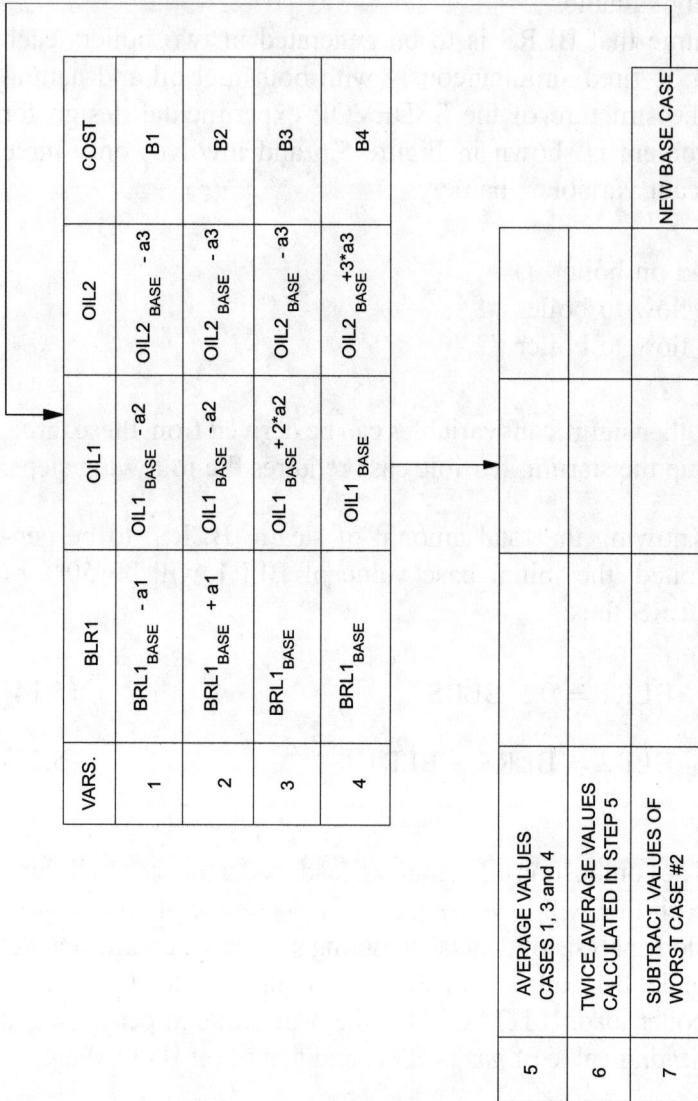

Figure 5.6   SSDEVOP experimental design - Two multi-fuel fired boilers

$$OIL2 = 0.5^*BLR2^*HTACQU/$$

$$(0.95^*HVO^*f(EFFO2(BLR2))) \qquad (5.17)$$

c. To calculate the total cost of the base case, it is necessary to calculate the gas flow needed to generate the balance of 50% of the boiler load, thus:

$$GAS1 = 0.5^*BLR1^*HTACQU/$$

$$(0.95^*HVG^*f(EFFG1(BLR1))) \qquad (5.18)$$

and $GAS2 = 0.5^*BLR2^*HTACQU/$

$$(0.95^*HVG^*f(EFFG2(BLR2))) \qquad (5.19)$$

from which, if $OIL and $GAS are fuel prices in $/MBTU, then:

$$TOTCST = ((OIL1 + OIL2)^*HVO^*OIL/1.0E + 06) +$$

$$((GAS1 + GAS2)^*HVG^*GAS/1.0E + 06) \qquad (5.20)$$

d. The initial base values for the three variables of the experimental design are now set to:

$$BLR1_{BASE} = BLR1$$

$$OIL1_{BASE} = OIL1$$

$$OIL2_{BASE} = OIL2$$

e. Initial values are now assigned to the perturbation parameters a1, a2 and a3. For these industrial boilers, the values were:

$$a1 = 2000lb/h$$

$$a2 = 200lb/h$$

$$a3 = 200lb/h$$

The experimental design array in Figure 5.6 may now be constructed from this data and the program executed. With the cost of oil substantially higher than that of natural gas, the optimum solution is shown in the first column of Table 5.5. The oil flow has been forced to its minimum value which would suggest that only natural gas should be fired to these boilers, given the fuel cost difference.

With the cost of oil only slightly higher than that of natural gas, the results are more interesting and shown in the second column of Table 5.5. Note that the optimal load distribution between the boilers has not changed significantly but advantage has been taken of the higher combustion efficiency of fuel oil and its reduced cost.

5.2.3   Optimization of VARS within a Small Refinery Distribution System

In addition to process steam and real power demands, manufacturing processes also have a varying reactive power demand that must be satisfied. Reactive power affects line currents and bus voltages, as well as the power factor of the tie-line bringing power into the plant from the local utility. Often the utility applies a penalty if the tie-line power demand exceeds an agreed value and frequently imposes a low power factor penalty as well. Stable system operation requires that bus voltages are maintained within assigned limits; that transformers and connecting cables do not become overloaded; and that generators and synchronous motors run within their reactive capabilities. There can be many acceptable combinations of the settings of generator and synchronous motor excitations, of transformer taps and capacitor banks, all of which lie within the set of electrical system constraints. However, there are so many variables to manipulate that it is very difficult for an operator to make optimal decisions, even under steady state conditions. It is an even more complicated

Table 5.   Optimal results for multi-fuel fired boilers

Parameter	Case #1	Case #2	Units
Cost of oil	5.6	4.7	$/MBTU
Cost of natural gas	4.3	4.3	$/MBTU
BLR1	141872	141566	lb/h
BLR2	162127	162433	lb/h
OIL1	300	3657	lb/h
OIL2	300	4057	lb/h
GAS1	134264	74266	scfh
GAS2	155259	8920	scfh
Total Cost	1333	1317	$/h

task during system upsets. Thus it is desirable that the operator have available a system that resolves the alternatives and presents the best combination of settings for implementation.

Such a system was provided to a power house for a small refinery in India that had three (3) combustion gas turbogenerators, each with its own heat recovery steam generator (HRSG), together with one steam turbogenerator arranged to form a combined cycle. All the HRSG's were equipped with both oil and gas burners for auxiliary firing. The steam from the HRSG's was not only the source of steam for the steam turbine but also supplied process steam demands VHP and HP. Steam extracted from the steam turbine supplied process steam demands MP and LP, pressure reducing valves also being included to augment the extraction flows if necessary. All of the generators supplied power at 11 Kv to the internal power distribution network, which was also connected to the local utility tie-line via tap-changing transformers. Many of the details of this system were outlined in Hanway and Putman (1992) and in Putman, Huff and Pal (1999).

A simplified steam/power distribution flow sheet for this plant is shown in Figure 5.7. The performance characteristic curves for all this equipment took the form of linear functions, even if some were discontinuous. As a result, linear programming could be used to optimize the assignments of power and steam flows among the turbogenerators and HRSG's; but the values for power contained in the solution are *real* power and take no account of the reactive power assignments needed to maintain bus voltages within tolerances, or tie-line power factor close to unity.

Note that power can be transferred from Line #1 to Line #2 and vice versa, depending on which turbogenerators are in operation and the relative loads on these two lines. For this reason, two variables were assigned to each of the two Lines (30 and 31 for Line #1 and 32 and 33 for Line #2), one for positive power flow and the other negative power flow to the Reference Bus. Meanwhile, tap-changing transformers were interposed between the reference bus and Lines #1 and #2 and the transformers have a maximum load limit. Thus variables 30 thru 33 are not only included in the power balance but are also constrained by the transformer load limit.

A simplified one-line electrical diagram for this plant is shown in Figure 5.8. The contract with the local utility included penalties not only for high power demands but also for low power factors on the utility bus. The distribution of VARS, or reactive power, within the network therefore became a requirement but there are at least seven equipment items that influence VAR distribution, namely:

1. Excitation of generator CGT#1
2. Excitation of generator CGT#2
3. Excitation of generator CGT#3
4. Excitation of generator SGT
5. Excitation of Synchronous Motor SYNCHM
6. Tap setting of tap-changing transformer TRANS#1
7. Tap setting of tap-changing transformer TRANS#2

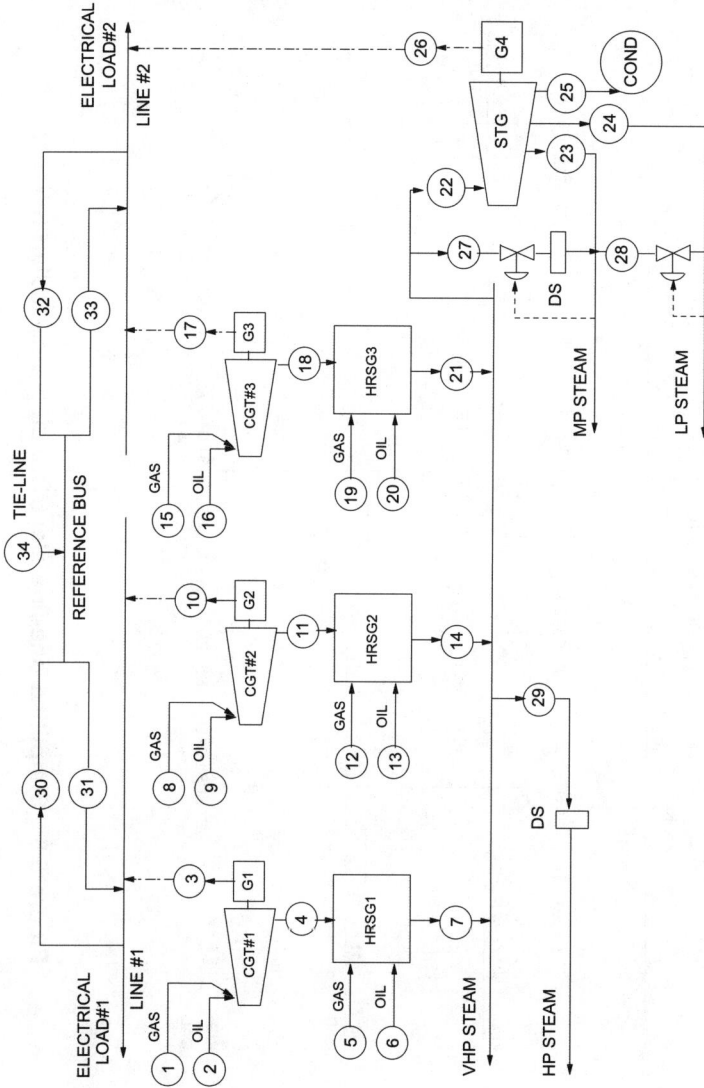

Figure 5.7   Steam/power distribution diagram - Indian refinery

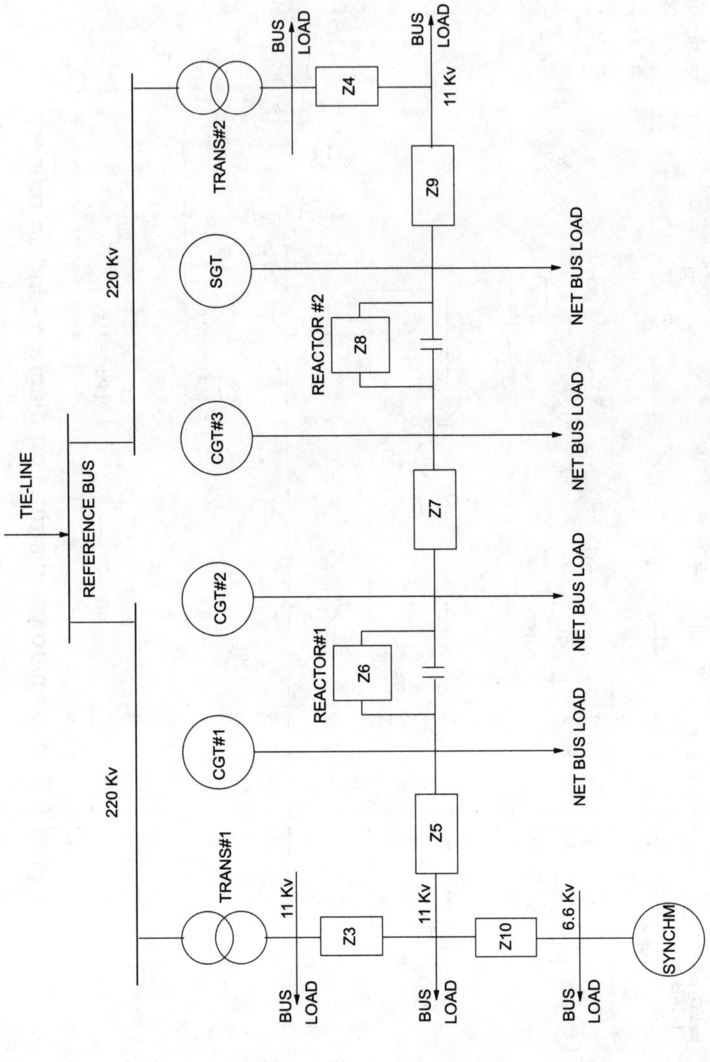

Figure 5.8 Simplified one-line electrical diagram - Indian refinery

The maximum capacity of the transformers was 33 MVA and their transformation ratios could be adjusted between 0.80 and 1.10 in 25 discrete steps. Two reactors, REACTOR#1 and REACTOR#2, were also included in the system but could be taken out of service if conditions warrant. The system control objectives were:

- Maintain the tie-line power factor as close a possible to unity
- Maintain the bus voltages within tolerances
- Maintain the assigned values of real power as generated by the LP program
- Ensure that the tap-changing transformers operate within their load and transformation ranges
- Ensure that all machines operate within their reactive capability curves (See Figure 6.4)

However, with so many control devices to choose from, many of which interact with each other, it is very difficult even for a skilled operator to determine exactly how to adjust all these devices to achieve the combined set of control objectives. It was therefore decided to provide the operator with an off-line program that could evaluate the choices with respect to some cost criterion and advise the operator of the recommended set of adjustments to make. The SSDEVOP technique was felt to be the most suitable for this purpose, using a seven-variable experimental design to search for the optimal settings for a given set of conditions. The ease of adapting the SSDEVOP algorithm for use with complex arithmetic was also a consideration.

Since the active and reactive power loops are essentially decoupled, it is possible to establish the optimal set of real power assignments using the linear programming technique but apply a different technique (i.e. SSDEVOP) for optimizing the VARS. The model that is perturbed by the SSDEVOP algorithm is a network constructed to match the present configuration of the plant equipment, as outlined by Elgerd (1971). The nodes of the network are the buses, which are connected together by lines. Each line has a specified impedance and loads are assigned to the buses as complex numbers indicating the amount of real and reactive power. Generators and synchronous motor loads are also assigned to the buses, again as complex numbers. In constructing the network to match its present configuration, circuit breakers are checked and, when open, a very high impedance is assigned to that line. In the case of tap-changing transformers, a multiplier is used to reflect the transformation ratio. The tie-line bus is assigned as the Reference Bus.

The model that is used to evaluate the network is a Static Load Flow Analysis (SLFA) program that uses the Newton-Raphson algorithm to solve a set of non-linear simultaneous equations that represent the admittance matrix for the electrical network. The static load flow equation for each bus takes the form:

$$P_i - jQ_i - y_{i1}V_1V_i^* - y_{i2}V_2V_i^* - \ldots - y_{iN}V_NV_i^* = 0$$

$$for\ i = 1, 2 \ldots N \tag{5.20}$$

Where

N = Total number of buses
P_i = Real power on bus i
Q_i = Reactive power on bus i
y_{ij} = Impedance between bus i and bus j
V_j = Voltage on bus j
V_i^* = Conjugate of voltage on bus i

It should be clear that there will be as many equations in the admittance matrix as there are lines connecting buses.

Referring to Figure 5.9, this matrix is constructed from the network details, the given data also including the net real power drawn from each bus. The model is then initialized with the present network values, including the reactive power on each bus and the SLFA executed. The variables that are adjusted to achieve convergence are the reactive power level at the reference (tie-line) bus, the reactive power setting on each machine, and the transformer transformation ratios that are required to maintain the bus voltages within their toler-ances, the voltage tolerances and reactive capabilities of all machines also being observed. This initial solution provides one set of reactive power and voltage settings corresponding to the set of real power flows on each bus. To establish the cost of this initial solution, weighting factors are applied to reference bus power factor, to bus voltage deviations and to machine reactive capability curve deviations and the total cost of the initial solution computed. This is the cost of case #8 in the 7-variable SSDEVOP tableau constructed for this plant, the form of Figure 5.6 being expanded to include seven rather than three variables.

Examples of weights are:

Tie-line power factor deviation from unity   10.0
Bus voltage deviation from 11 Kv   0.1
Reactive capability violation   1.0

Referring to Figure 5.1, for each case #1 thru #7, the reactive power component in each of the variables 1 thru 7 defined above is now perturbed around the base case in accordance with the SSDEVOP tableau. Typical perturbation values were 2.0 VARS and 0.05 for the tap changing trans-former multipliers. The model was then allowed to converge on the set of bus voltages and tie-line power factor that satisfy the corresponding set of equations, with not only the real

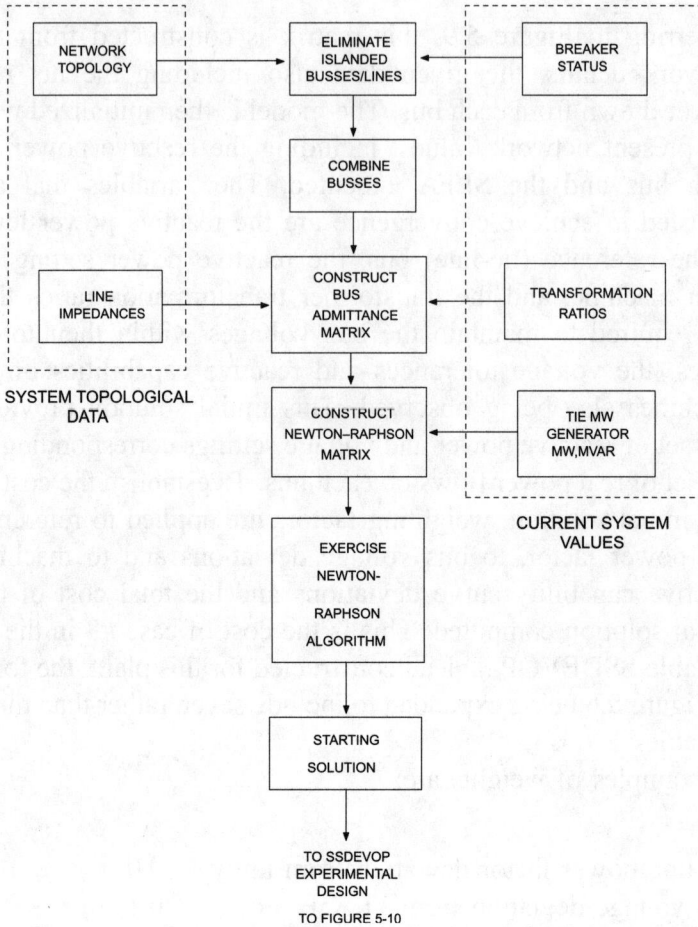

```
┌ ─ ─ ─ ─ ─ ─ ─ ─ ─ ─ ─ ─ ─ ─ ─ ─ ─ ─ ─ ─ ─ ┐
              ELIMINATE
   NETWORK    ISLANDED          BREAKER
   TOPOLOGY   BUSSES/LINES      STATUS

              COMBINE
              BUSSES

   LINE       CONSTRUCT
   IMPEDANCES BUS               TRANSFORMATION
              ADMITTANCE        RATIOS
              MATRIX
─ ─ ─ ─ ─ ─ ─ ─ ─ ─ ─ ─ ─ ─ ─ ─ ─ ─ ─ ─ ─ ─
SYSTEM TOPOLOGICAL
      DATA     CONSTRUCT        TIE MW
               NEWTON–RAPHSON   GENERATOR
               MATRIX           MW,MVAR

                                CURRENT SYSTEM
               EXERCISE         VALUES
               NEWTON-
               RAPHSON
               ALGORITHM

               STARTING
               SOLUTION

               TO SSDEVOP
               EXPERIMENTAL
               DESIGN
               TO FIGURE 5-10
```

Figure 5.9   Matrix construction and SSDEVOP initialization

powers but also these reactive power values fixed. After convergence, the results of each case may now be costed using the established set of weighting factors and a new base case established using the standard SSDEVOP procedure. The new set of experiments is now evaluated and the costs are compared, the SSDEVOP algorithm calculating the set of controlled variables that will move the system to a more optimum state, by giving predominance to those cases where

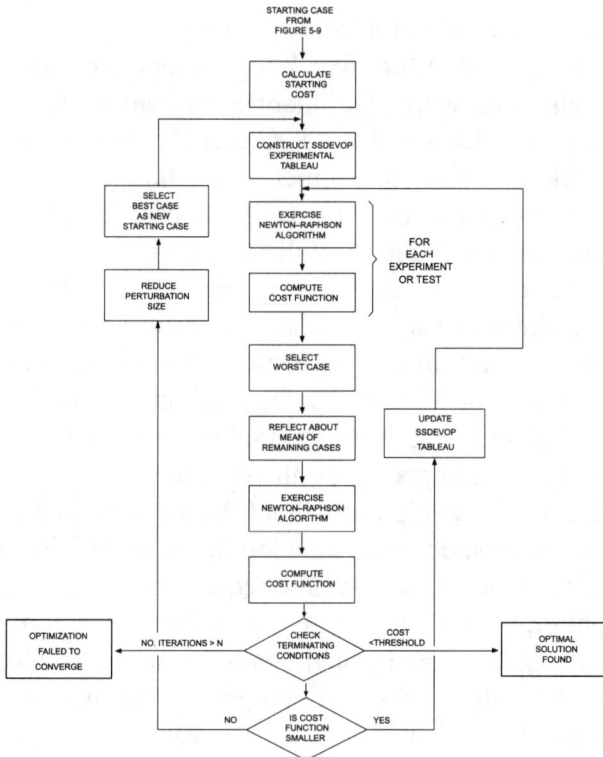

Figure 5.10   SSDEVOP experimental design optimizing procedure

the set of controlled variables has the least weighted cost. The process terminates when the new base cost is greater than its previous value; or the actual cost, or the reduction in base cost, are less than predetermined amounts; or the number of iterations exceeds a predetermined value. At this point, and assuming convergence, it is possible to decrease the perturbation size and repeat the process until a new and more accurate convergence is obtained. The final set of perturbed variables are those which should be transferred to the machines in order that all the plant variables, particularly bus voltages and tie-line power factor, should be at their optimal settings.

The off-line program allows a variety of operating conditions to be evaluated, the weighting factors and convergence criteria adjusted, while the manner in which the program converges can also be studied. Clearly, convergence should occur within a reasonable number of iterations and appropriate adjustments are made to the convergence criteria to ensure this. Arrangements can be made to allow the program to run on an operator request, or automatically whenever a load or bus voltage deviates outside a preset deadband. An on-line program can also be provided, executed after a stated amount of time has elapsed or whenever a deadband is exceeded, arrangements being made to transfer the optimal settings to the equipment after acceptance by the operator.

Ideally, all the equipment should always be available for control. Unfortunately, this is seldom the case and the program must reflect whether an item of equipment is available for VAR control, is on manual or is out of service. If some of the equipment and the associated loads become islanded, it is possible that only a subset of equipment and their associated network can be evaluated.

5.3   Equal Incremental Cost Applications

The equal incremental cost method for optimizing the load distribution among a set of similar equipment with non-linear characteristics was introduced in Section 4.4. The set of equipment must be delivering a common energy resource (e.g. power in the case of turbogenerators or steam at the same conditions in the case of boilers) and the cost of that energy must be derivable from a function that is dependent on the amount of the energy resource being delivered. Thus turbogenerators would be provided with a heat rate vs. load (kw) curve as well as the cost of fuel in $/MBTU. Boilers would be provided with an efficiency vs. load (lb/h) curve as well as the cost of the fuel being fired to that boiler. Note that each boiler must be fired with only one fuel but it does not have to be the same fuel on each boiler. Hydro-electric generators would be provided with a curve showing the relationship between water flow and generator load (kw). Suitable curves would be provided for other equipment such as chillers.

Once the basic characteristic curves have been identified, the principles outlined in Section 4.4 can be used to optimize the distribution of load among the set of equipment, the results being applied as appropriate after they have been studied. Clearly, the sum of the initial set of load assignments must be equal to the total load to be optimized. In the case of a program to be used for on-line optimization, the initial loads would be the present set of loads at which the equipment is operating. In the case of an off-line program, the user would determine the total load (L) to be studied as well as the number and identity of the units (N) to be included in the optimized set, the initial loads then being set to L/N. Such off-line studies allow a wide range of operating conditions to be explored without disturbing the plant and the lessons learned can be incorporated in the on-line program. Experience has shown that the cost of optimized results must always be either equal to or better than that of the present case. If the optimized results are more costly, something is wrong with the data.

5.3.1 Range of Boilers

Consider a range of four (4) identical boilers, each fired with the same fuel, having a maximum capacity of 350 klb/h and with the point of inflexion on the total heat input curve being at 234.9 klb/h. The allowable operating range was between 80 and 350 klb/h and Table 5.3.1 shows the results calculated for a variety of different total loads, the initial cost being for the case when all boilers were intuitively assigned the initial values of L/N.

Tests 1X thru 10X show the results when all four boilers remain in the optimizing set. When the total loads are relatively low, some of the boilers are driven down to their minimum loads; they probably should not be in operation under those conditions unless management knows that the load is to be increased shortly. Even so, it is seen that the cost of the optimal assignments is less than the intuitive initial assignments. However, the savings diminish as the total load increases, to the point where the savings fall to zero when the value of L/N is well above the point of inflexion. Where units have been forced to the low end of their range, shutting them down (if appropriate) can also realize substantial savings, as shown in the results of tests 1Y thru 10Y in Table 5.3.1.

The results shown in table 5.3.1 are the general experience. However, this table assumed boilers with identical characteristics. This is seldom the case in practice and variations in fuel or differences in boiler construction often provide enough variations to make the optimized results even more interesting.

5.3.2   Hydroelectric Turbogenerators

Two dams on the Columbia River in Oregon had eight (8) hydroelectric turbogenerators each, the generators on each dam being originally dispatched as a group by the central dispatcher. The need to conserve water and to facilitate the passage of fish upstream during the spawning season caused new strategies to be developed.

Strategy (A) applied during much of the year. With this strategy, the generators on each dam were still dispatched as a group but the dispatcher included the availability of water in the calculations. Once the total amount of power to be generated was determined, it was the responsibility of the

Table 5.3.1   Optimal boiler load distributions

Test	Total klb/h	Boiler #1	Boiler #2	Boiler #3	Boiler #4	Initial Cost $/h	Savings $/h	%
1X	560	80	320	80	80	1139.8	44.7	3.92
1Y		0	0	280	280	1139.8	152.9	13.4
2X	640	80	255	80	225	1267.3	38.95	3.07
2Y		0	0	320	320	1267.3	154.6	12.2
3X	720	80	280	80	280	1387.3	41.23	2.97
3Y		0	80	320	320	1387.3	247.3	17.82
4X	800	80	320	80	320	1505.2	33.31	2.21
4Y		0	0	400	400	1505.2	96.02	6.37
5X	880	85	265	265	265	1622.1	19.8	1.22
5Y		0	80	400	400	1622.1	185.6	11.44
6X	960	225	255	225	255	1739	0.559	0.03
6Y		0	320	320	320	1739	69.9	4.01
7X	1040	260	260	260	260	1854.9	0	0.0
7Y		0	345	345	350	1854.9	49.54	2.67
8X	1120	280	280	280	280	1973.8	0	0
8Y		280	280	280	280	1973.8	0	0
9X	1200	300	300	300	300	2096.9	0	0
9Y		300	300	300	300	2096.9	0	0
10X	1280	320	320	320	320	2225.4	0	0
10Y		320	320	320	320	2225.4	0	0

dam to *minimize* the amount of water consumed to generate that power.

This is a classical non-linear problem, the best solution for which uses the equal incremental cost principle. Power was the energy resource being delivered, the characteristic curve for each hydro-electric turbine being stated in terms of water flow vs. generated power. It is interesting to note that some ranges of power were excluded because of vibration problems being experienced when operating within these ranges and the optimization programs had to take these prohibitions into account. Clearly, the total amount of power required by the central dispatcher would determine how many units had to be in operation but the units selected would be those that would minimize the total amount of water consumed. It was the task of the off-line optimization program to make this selection.

Strategy (B) applied during the spawning season. In this case, the amount of water allowed through the turbines was determined and it was the task of the operating staff to maximize the amount of power this could generate. Again, the equal incremental cost principle was used but, in this case, the common resource was water flow, with the dependent variable being power. Thus the turbine characteristic curves had to be interpreted in a different way but the structure of the optimization program for strategy (B) was similar to that of strategy (A). Clearly, the prohibited zones still had to be avoided.

Because of the much lower water flow, only some of the turbines would be selected for operation. These would tend to be the most efficient units but an off-line version of the program was needed to make the selection.

5.3.3   In-plant Dispatching of Steam Turbogenerators

The economic dispatching of utility generator units is normally performed by applying a LaGrange multiplier to the incremental cost curves, often defined for each unit as a discontinuous sequence of linear relationships. Any errors introduced by using these linear approximations are comparatively small for the system as a whole. However, when dispatching a group of steam turbogenerators located within one plant, greater accuracy can be obtained by representing the heat rate and incremental cost curves as continuous polynomials, the coefficients in which are generated from historical plant data using regression analysis.

A typical heat rate vs. load curve is shown in Figure 5.11 and can be described by a polynomial having the following form:

Let MW = generated load MW
   HR = unit heat rate BTU/Kwhß
    β = boiler combustion efficiency %
TOTHT = total heat acquired BTU/h
 INCST = incremental cost BTU/Kwh
   b0, b1, b2 and b3 are constants

Then

$$HR = 1/\beta {}^{*}(b_0 + b_1{}^{*}MW + b_2{}^{*}MW^2 + b_3{}^{*}MW^3) \qquad (5.21)$$

Note that heat rate is an inverse measure of turbo-generator efficiency ε, such that:

$$\varepsilon = 3600/HR \qquad (5.22)$$

The total heat vs. load relationship can be obtained by multiplying Equation (5.21) by MW * 1000, thus

$$TOTHT = 1000/\beta {}^{*}(b_0 MW + b_1{}^{*}MW^2$$

$$+ b_2{}^{*}MW^3 + b_3{}^{*}MW^4) \qquad (5.23)$$

or, converting MW to Kw

$$TOTHT = 1/\beta {}^{*}(b_0 Kw + b_1{}^{*}1.0e - 03 {}^{*}Kw^2$$

$$+ b_2{}^{*}1.0e - 06 {}^{*}Kw^3 + b_3{}^{*}1.0e - 09 {}^{*}Kw^4) \qquad (5.24)$$

Incremental cost can be obtained by differentiating (5.24), thus:

$$INCST = 1/\beta {}^{*}(b_0 + 2 {}^{*}b_1{}^{*}1.0e - 03 {}^{*}Kw +$$

$$3 {}^{*}b_2{}^{*}1.0e - 06 {}^{*}Kw^2 + 4 {}^{*}b_3{}^{*}1.0e - 09 {}^{*}Kw^3) \qquad (5.25)$$

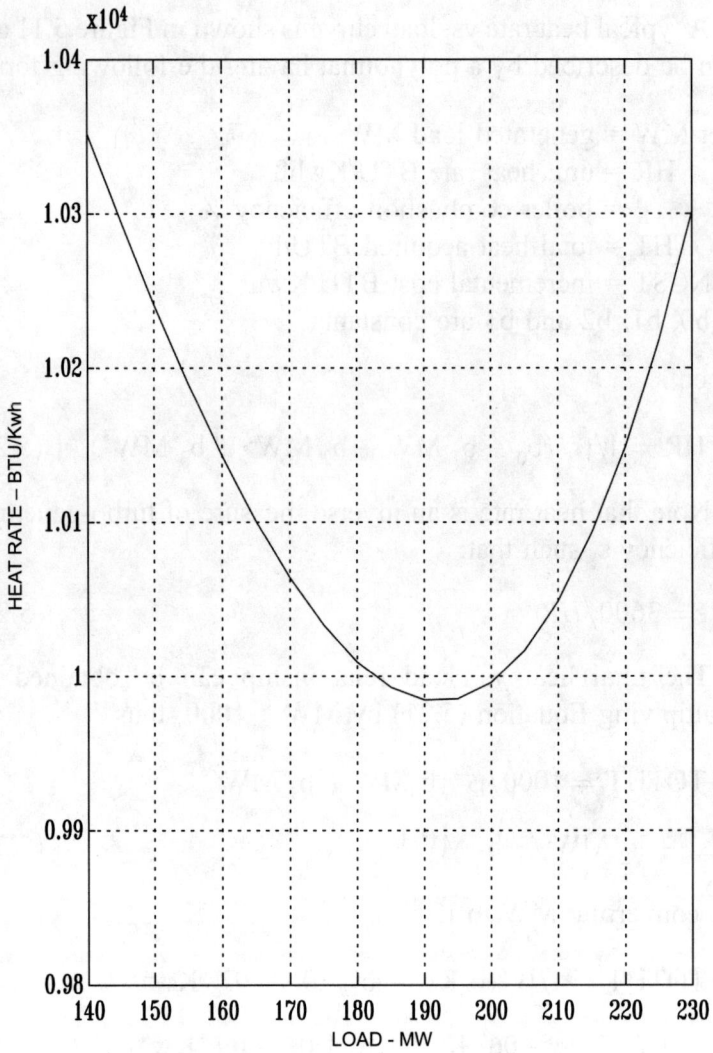

Figure 5.11   Heat rate vs. load

It is important to note that the constants in Equation (5.25) for incremental cost are not the same as those in Equation (5.21) for heat rate. As Putman (1978) pointed out, incremental cost is not the same as heat rate, although the curves have similar shapes.

Figure 5.12 shows total heat and incremental cost plotted vs. load, based on the turbogenerator heat rate shown in Figure 5.11. The unit heat rate is obtained by dividing the turbogenerator heat rate by boiler combustion efficiency. The curves in Figure 5.12 show the relationships for boiler efficiencies of 100%, 95% and 90% respectively and indicate how boiler efficiency affects both total heat and incremental cost for a given unit. This can have a significant affect on the optimal load distribution and it should be clear that the current boiler efficiency must be taken into account within the economic dispatch program. A typical program flow sheet to generate and update the data base associated with in-plant unit economic dispatching is shown in Figure 5.13. Meanwhile, the dispatching procedure itself closely follows that outlined in Section 5.3.1 above.

5.3.4 Optimal Load Trajectories

The in-plant dispatching of a group of steam turbogenerators is frequently practiced in European power plants. However, the total plant load requested by the central dispatcher can change over short time periods which requires recalculation of the set of optimum loads to be applied to the units. To reduce the amount of calculation to be performed before the total load transition occurs, it is possible to calculate the optimal trajectory that each unit should follow during an anticipated load transition.

It is clearly possible to calculate, regress and plot the unit load versus the corresponding incremental cost and to do this individually for all the units in the plant. Equation (5.25) states

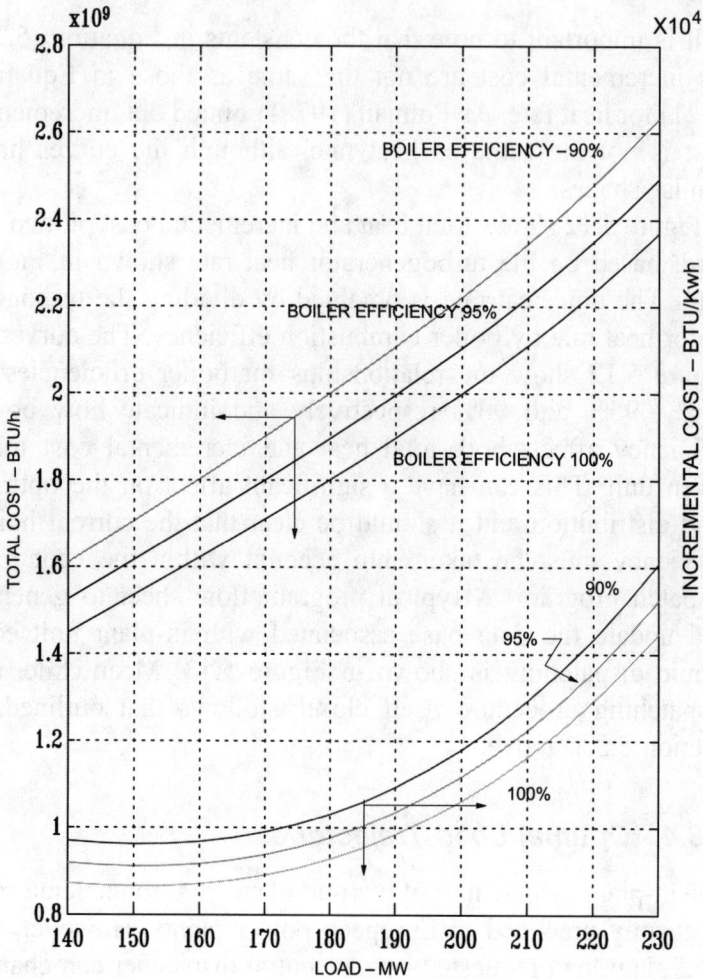

Figure 5.12   Total cost and incremental cost vs. load

the relationship between incremental cost and load in the form
of a third-order polynomial. However, this relationship can be
transformed into that between incremental cost IC and the unit
load MW, having the form:

$$MW = a_0 + a_1 IC + a_2 IC^2 + a_3 IC^3 \qquad (5.26)$$

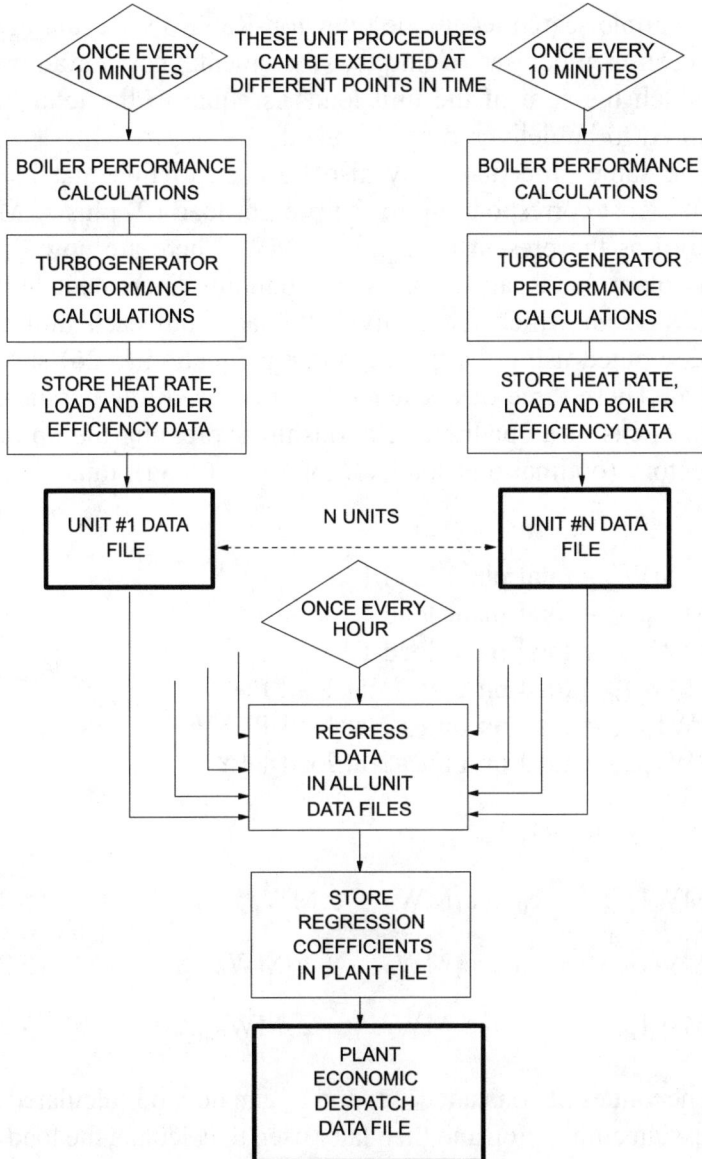

Figure 5.13   Data base generation subsystem

A simple search technique (Newton-Raphson, Regula Falsi, etc.) can then be used to find that incremental cost for all units at which the sum of the unit loads is equal to the total load required to be delivered by the plant.

The same procedure may also be used to find that incremental cost corresponding to the present load (P) plus X MW as well as the present load plus Y MW. There are now three total plant loads and three corresponding incremental cost values, from which the equivalent load from each unit can be computed using the corresponding Equation (5.26) shown above. These three data sets can now be used to compute the coefficients in a quadratic relationship expressing the optimal trajectory (optimal unit load vs. total plant load), thus:

if $\quad MW_P$ = total plant load P
$\quad MW_{P+X}$ = total plant load P+X
$\quad MW_{P+Y}$ = total plant load P+Y
$\quad MW1_P$ = load on unit #1 at load P
$\quad MW1_{P+X}$ = load on unit #1 at load P+X
$\quad MW1_{P+Y}$ = load on unit #1 at load P+Y

Then, for unit #1,

$$MW1_P \quad = \quad c_0 + c_1 MW_P + c_2 MW_P^2 \qquad (5.27)$$

$$MW1_{P+X} = \quad c_0 + c_1 MW_{P-X} + c_2 MW_{P-X^2} \qquad (5.28)$$

$$MW1_{P+Y} = \quad c_0 + c_1 MW_{P+Y} + c_2 MW_{P+Y^2} \qquad (5.29)$$

The values of constant c_0, c_1 and c_2 can now be calculated by Gaussian elimination and then later used to calculate the load on unit #1 as a function of the total plant load. The procedure would be repeated for units 2 thru N. Figure 5.14 shows a typical plot of the optimal trajectories for each of several units, in terms of the optimal load for each unit vs. the total plant load.

It is important that time be spent analyzing off-line the behavior of the plant, the performance of which it is planned to optimize. Figure 5.15 shows the groups of data that would be needed and one suggestion for an operator interface. Meanwhile, Figure 5.12 indicates that there may be two loads with the same incremental cost at lower loads. Clearly, over the load ranges X and Y that are used to generate Equations (5.27–5.29), it is not admissible for the optimal assignments to switch from the lower to the higher load even if the total cost could be reduced by doing so. Thus, for any given unit and value of P, the optimizing algorithm has to include means for using only the higher or the lower of these two loads to generate the data for equations (5.27–5.29).

Once the optimizing policy has been determined for each of a wide range of operating conditions, the on-line program can be finalized. On-line control will usually be performed within a distributed process control system and, once the values of c_0, c_1 and c_2 have been determined they should be downloaded to the appropriate location within the control system, as indicated in Figure 5.16. Subsequently, as the value of MW changes, the corresponding optimal load assignment for each unit on-line will be calculated using these coefficients.

5.4 Combined Cycle Optimization

In combined cycle plants, while the models for much of the equipment, and for energy and mass balances, can be defined as linear equations, including that for the steam turbogenerator (STG); the models for the gas turbogenerators (GTG's) and heat recovery steam generators (HRSG's) are inherently non-linear. Meanwhile, there are a number of equipment design and operating constraints which must be observed; and frequently, the operating conditions have to be adjusted to ensure that emission constraints are not violated. It is also clearly

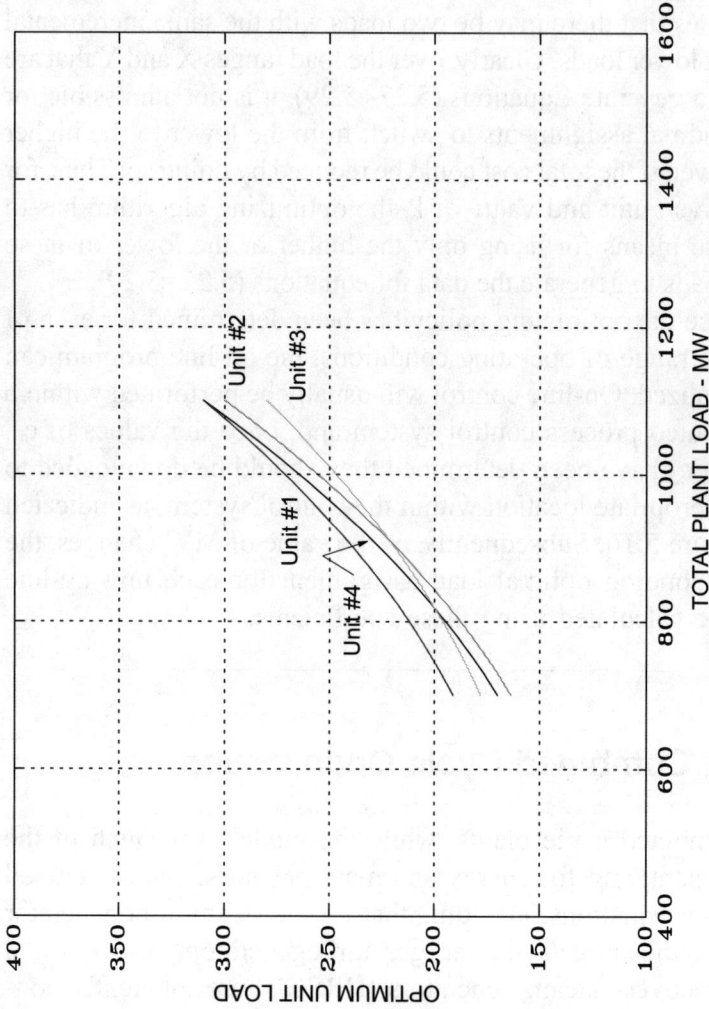

Figure 5.14   Optimal trajectories

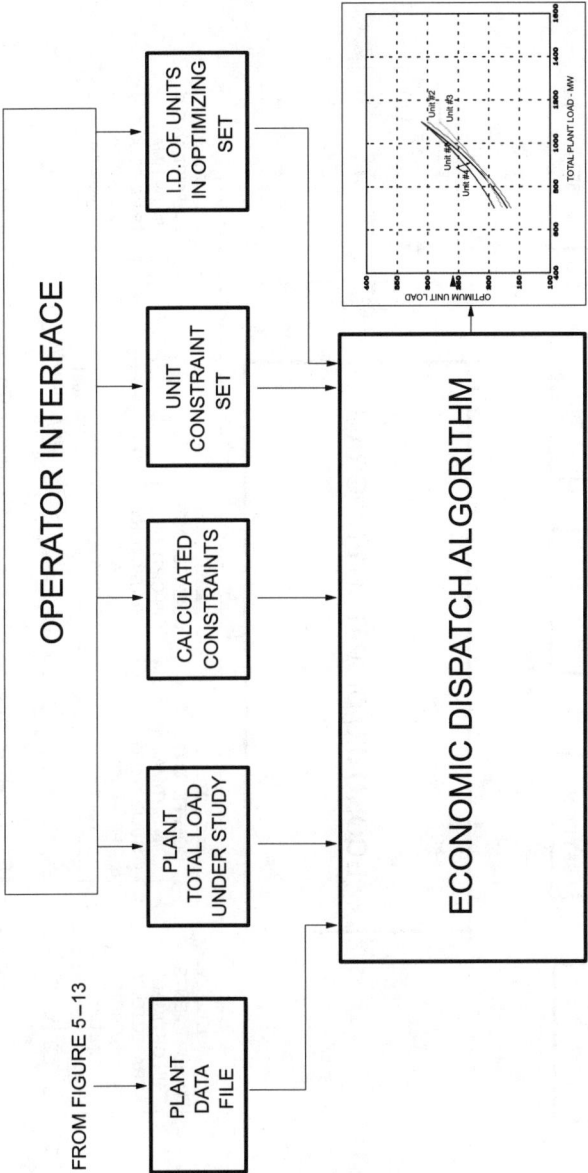

Figure 5.15   Economic dispatch program and interfaces

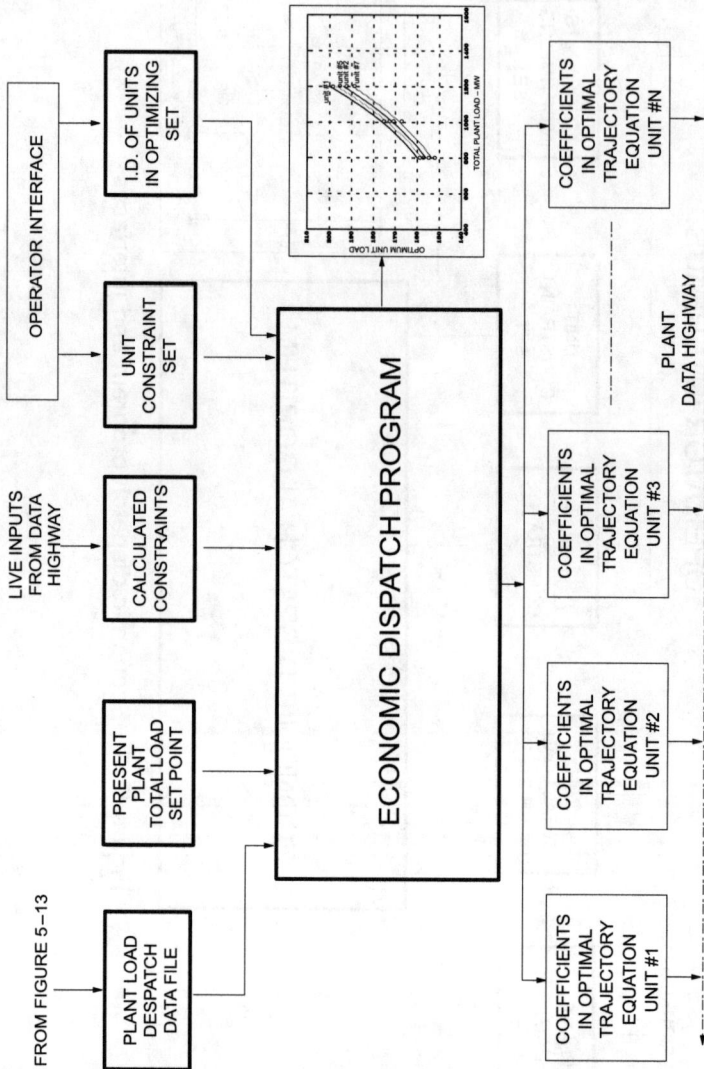

Figure 5.16   On-line economic dispatch program and interfaces

desirable that the final set of operating conditions should allow the combined demands for both generated power and process steam (where appropriate) to be met in such a way that revenue is also being maximized, or the operating cost minimized.

The steady state modeling of combined cycle plants has been discussed by several authors, including Horlock (1992) and Kehlhofer (1991) with special emphasis on plant design. Ordys et al. (1994) discussed the dynamic modeling of these plants and Katebi et al. (1995) focused on control problems, available software tools, and the optimization of, for example, dynamic set point maneuvers. Meanwhile, the on-line economic optimization of combined cycle plants has seldom been addressed and Putman et al. (1996) show that this is a topic with potentially rewarding results both at the plant design stage and during plant operation.

Putman et al. (1992) showed how the linear programming algorithm could be used to optimize a plant model even though it contains a number of non-linear relationships. The method used was to perturb the non-linear models about the last significant solution point; linearize them and update the appropriate coefficients in the L.P. matrix; and then run the L.P. algorithm again. The problem is defined as having converged when the total cost of the current solution is sensibly the same as the total cost of the previous solution. The major requirement as stated by Peressini (1988) is that the non-linear models be continuous and convex, the definition of the latter being when:

$$f(x) = f(x_0) + \frac{\partial f(x_0)}{\partial x} (x - x_0) \tag{5.30}$$

5.4.1.1 *Data Flow*

The discussion will center on the typical combined cycle plant flow sheet shown in Figure 5.17. This plant consists of a

42.5 MW gas turbogenerator (GTG) in which the Nox emissions are controlled by the injection of steam into the combustor. The exhaust gases are then passed to a heat recovery steam generator (HRSG) designed to raise steam at two different pressures (163 Klb/h at high pressure and 2 Klb/h at low pressure). Note that some modification to the following discussion might have to be made if the system configuration were to change, e.g. if duct burners were to be located between the GTG and the HRSG.

The high pressure steam is passed to a 15 MW extracting/ condensing type steam turbogenerator (STG) from which steam is extracted at a pressure required by the adjacent process plant. As a backup, the process steam can be supplied, or augmented, by passing high pressure steam through a pressure reducing valve (node 16) and desuperheating it. The Nox steam source is also extracted from the STG or can be taken from the high pressure steam supply through a pressure reducing valve (node 14) and desuperheated.

The low pressure steam is the source of heat to feed heaters, any surplus being *inducted* into one of the low pressure stages of the STG.

Note that Nox steam not only decreases the Nox emissions but also increases the capacity of the gas turbogenerator due to the increased density of the gases passing to the turbine. However, the extraction of Nox steam from the turbine can also reduce the capacity of the steam turbogenerator. Clearly, there are a number of interactions between various equipment items; and these have to be reflected in the set of linear equations which form the LP matrix.

For a combined cycle plant which is not equipped with duct burners, the essential data flow is shown in Figure 5.18. When in a state of equilibrium, all the conditions around a gas turbogenerator may be considered to be a function of generated power and ambient air temperature. These include the exhaust gas flow and temperature, as shown in Figure 5.19. Meanwhile

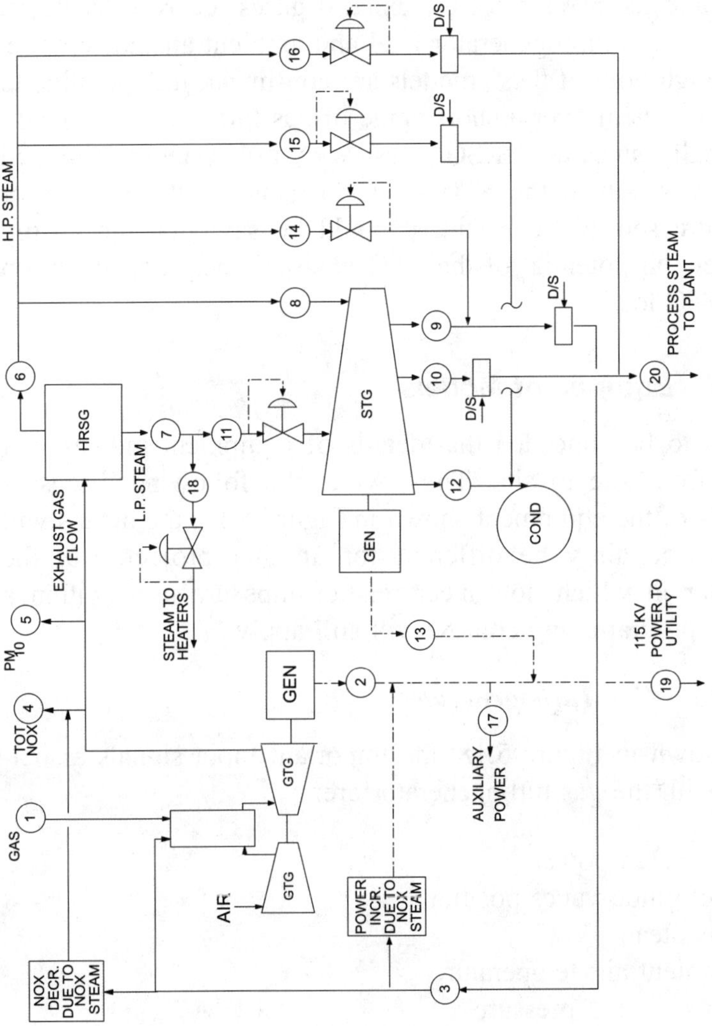

Figure 5.17   Combined cycle and cogeneration flow sheet

the performance of a two-pressure HRSG, especially as reflected in the mass flow rate and temperature of the HP steam and the flow of LP steam, is largely a function of the flow and temperature of the exhaust gases received by it, i.e. again of gas turbogenerator load and ambient air temperature. Although both of these models are non-linear, it is possible to linearize them over a short range of gas turbogenerator load.

Finally, since the HRSG pressure control system is designed to pass steam to the STG so as to maintain the HP header pressure sensibly constant, it could be said that the energy conversion potential of the STG is also largely dependent on the GTG load.

5.4.2   Equipment Models

It has to be conceded that details of equipment models will vary from site to site. Thus, while the following discusses details of the equipment shown in Figure 5.17, the details will almost certainly be different for another project. But the manner in which non-linear relationships are used within a linear programming context will still apply.

5.4.2.1   Gas Turbogenerator

As shown in Figure 5.18, the important input signals associated with the gas turbogenerator are:

Generated power
Inlet guide vanes position
Nox steam flow
Ambient air: temperature
              pressure
              humidity

These signals, together with the gas turbogenerator model, determine the steady state performance of the gas

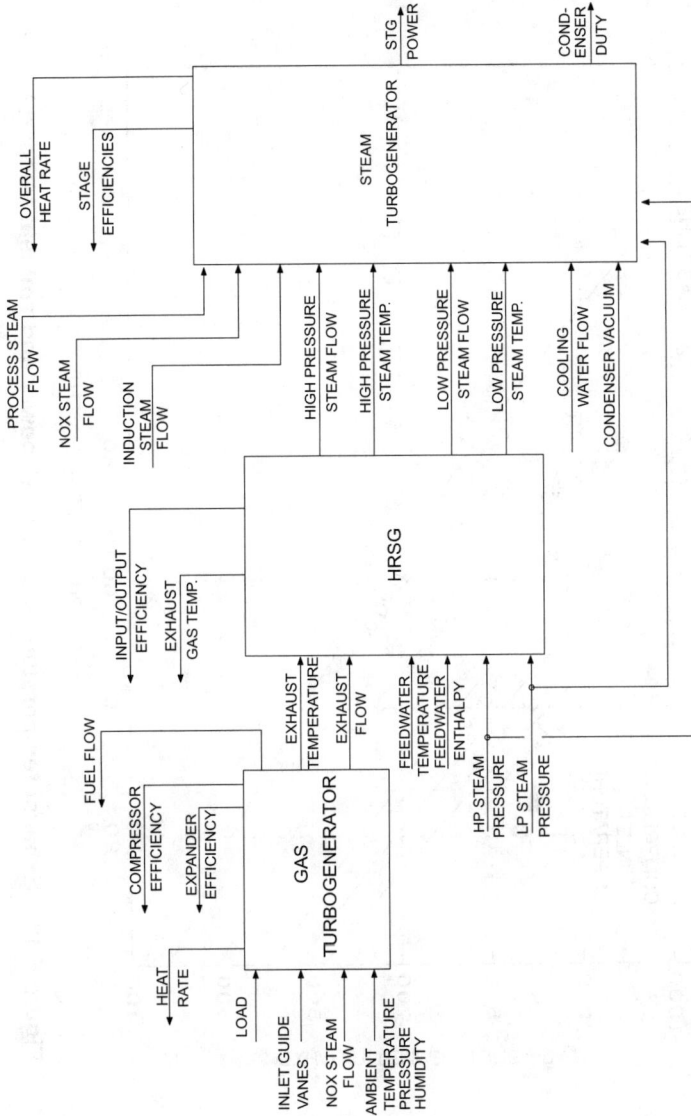

Figure 5.18   Combined cycle performance calculations information flow sequence

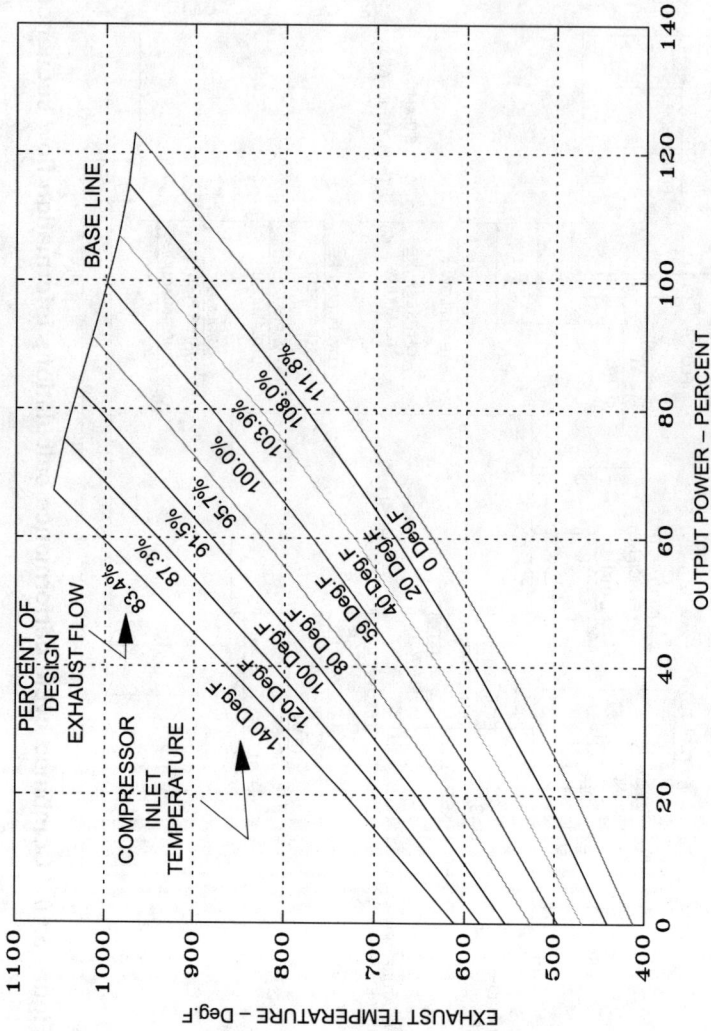

Figure 5.19   Exhaust temperature and flow vs. power and ambient temperature

turbogenerator, in terms of fuel flow rate, heat rate, compressor and expander efficiencies, together with the quantity and temperature of the exhaust gases.

Let: GF = Fuel Gas flow   MMBTU/h
   MWGT = Gas Turbine power MW
  MWGTMX = Maximum gas turbine power MW
      AMB = Ambient air temperature   Deg.F
      EXF = Exhaust flow from GTG   Klb/h
    EXHT = Exhaust gas temperature   Deg.F
   NOXF = Nox steam flow   Klb/h

Then:

$$GF \qquad\qquad = f_1(MWGT, AMB) \qquad\qquad (5.31)$$

$$MWGTMX \qquad = f_2(AMB) \qquad\qquad (5.32)$$

$$EXF \qquad\qquad = f_3(AMB) \qquad\qquad (5.33)$$

$$EXHT \qquad\qquad = f_4(MWGT, AMB) \qquad\qquad (5.34)$$

$$NOXF \qquad\qquad = f_5(MWGT, GF) \qquad\qquad (5.35)$$

Prior to the LP execution, the current value of MWGT and AMB are presented to a subroutine which calculates the parameters defined in Equations (5.31–5.35) above and stores the results; then 1 MW is added to MWGT and the same subroutine is used to calculate the new parameter set for (MWGT + 1.0). Using these results, a simple calculation allows the current values of the coefficients in the following equations (Y.1) and (Y.2) to be determined:

$$GF - a_1 * MWGT \qquad = a_0 \qquad\qquad (Y.1)$$

$$b1 * GF - b_2 * NOXF = b_0 \qquad\qquad (Y.2)$$

while

$$MWGT \leq MWGTMX \tag{Y.5}$$

5.4.2.2   Heat Recovery Steam Generator

Figure 5.2 is a schematic showing the major heat transfer surfaces in a typical two-pressure unfired Heat Recovery Steam Generator. The heat source is the exhaust gases from the gas turbogenerator, the significant parameters for which are the exhaust gas flow (EXF) and temperature (EXHT). The signals used by the HRSG model, in addition to these two, are:

> TFW = Feedwater temperature
> TFH = Feedwater enthalpy
> PHPSTM = High pressure steam pressure
> THPSTM = High pressure steam saturation temperature
> TSTM = High pressure steam temperature after desuperheating
> FSTM = High pressure steam flow after desuperheating
> TISTM = High pressure steam temperature before desuperheating (Superheater outlet temperature)
> FISTM = High pressure steam flow before desuperheating
> TSHOUT = Gas temperature at outlet of superheater pass
> THEVOUT = Gas temperature at outlet of H.P. economizer
> PLPSTM = Low pressure steam pressure
> TLPSTM = Low pressure steam temperature (Saturated steam temperature)
> FLPSTM = Low pressure steam flow
> TLEVOUT = Gas temperature at outlet of L.P. economizer
> UA1 = Heat transfer coefficient for superheater pass
> UA2 = Heat transfer coefficient for H.P. pass
> UA3 = Heat transfer coefficient for L.P. pass

Note that the values of UA1 through UA3 are the effective heat transfer coefficient multiplied by the corresponding heat transfer area and have the units BTU/h.Deg.F.

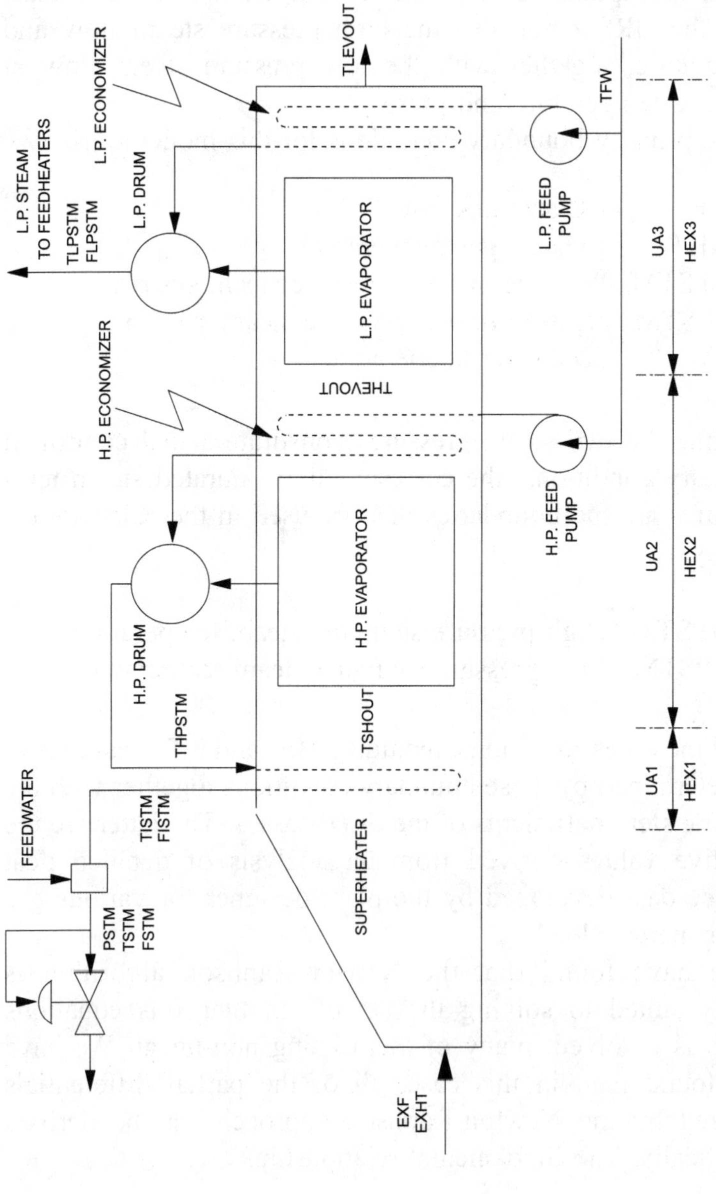

Figure 5.20   Heat recovery steam generator heat transfer surfaces

The steady state model of the HRSG not only calculates the input/output efficiency and the temperature of the exhaust from the HRSG but also the high pressure steam flow and temperature, together with the low pressure steam flow at saturated temperature and pressure.

The primary boundary conditions for this model are:

EXF   Exhaust gas flow
EXHT   Exhaust gas temperature
PHPSTM   Pressure of high pressure steam system
PLPSTM   Pressure of low pressure steam system
TFW   Feedwater temperature

While the two steam pressures constitute actual controlled boundary conditions, the corresponding saturated steam temperatures are the boundaries that are used in the calculations, namely:

THPSTM   High pressure saturated steam temperature
TLPSTM   Low pressure saturated steam temperature

All the other conditions, including H.P. and L.P. steam flows are determined by these boundary conditions together with the heat transfer coefficients of the three passes. The latter are the effective values derived from an analysis of detailed heat balance data developed by the plant designer for various gas turbogenerator loads.

We have found that the Newton-Raphson algorithm is ideally suited to solving the set of simultaneous equations which is involved, many of them being non-linear. We have also found that, in this case, all of the partial differentials required by the Newton-Raphson approach can be derived analytically. The fundamental relationships are:

A. Heat transferred to steam
   as a function of temperatures = heat lost by flue gases
   and heat transfer coefficient
B. Flow of steam multiplied by = heat lost by flue gases
   enthalpy rise

In the case of the superheater, equation A takes the form:

$$UA1 * \frac{(EXHT - TISTM) - (TSHOUT - THPSTM)}{\log \frac{(EXHT-TISTM)}{(TSHOUT-THPSTM)}}$$

$$= EXF\,{}^*(EXHT - TSHOUT) \qquad\qquad (5.36)$$

while equation B takes the form:

$$FISTM\,{}^*(HISTM - HHPSTM)$$

$$= EXF^*(EXHT - TSHOUT) \qquad\qquad (5.37)$$

In the above two equations, the prefix H is used to indicate an enthalpy corresponding to the associated temperature.

5.4.2.3   Steam Turbogenerator

The characteristic curve for this steam turbogenerator is shown in Figure 5.21. Let:

T = throttle flow
HF = HP extraction flow
LF = LP extraction flow
P = Generated power
a_o = Intercept shown in Figure 5.21
a_1 = Incremental throttle flow for incremental power
   (Slope of Willans line)

a_2 = Incremental throttle flow for incremental LP flow
a_3 = Incremental throttle flow for incremental HP flow

$$T = a_0 + a_1 \cdot P + a_2 \cdot LF + a_3 \cdot HF \tag{5.38}$$

However, if the steam or vapor pressures should vary from their design values then adjustments will have to be made to the coefficients in the above equation.

5.4.3 Linear Programming Matrix

By linearizing the equipment models outlined above, the network connecting them can be defined within a linear programming matrix, in which are included not only the linear relationships inherent in the behavior of the plant steam/power equipment and distribution network but also the operating and design constraints. Prices within the cost function are assigned only to all externally purchased energy resources as well as to the cost of violating emission constraints. After linearizing the equipment models over a short GTG load range, the linear coefficients so calculated are inserted into the appropriate elements of the LP matrix. At the end of each execution of the LP, the equipment models are updated with the new linearized coefficients obtained by perturbing them about the new GTG load assignment contained in the previous solution. This iterative procedure continues until the value of the revenue calculated at the solution remains essentially constant between two iterations.

Among the constraints included in the right hand side column (RHS) of the LP matrix are the desired amount of process steam to be delivered to the plant as well as the amount of power that has been requested to be generated by the utility dispatch center. The optimum solution contains the values of the column variables which allow all of the defined constraints and equalities, together with the combined energy demands to

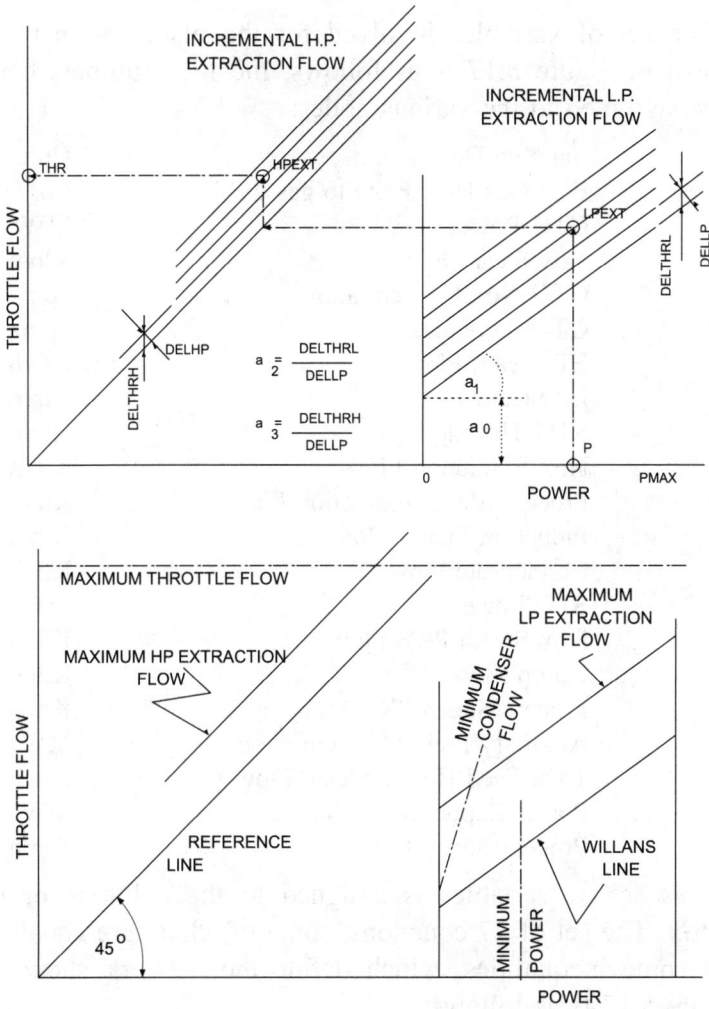

Figure 5.21   Typical turbogenerator characteristic curves

be satisfied, with the revenue being maximized at the same time. In certain situations infeasible solutions result, and these draw attention to a potential conflict between energy demands and system capability.

The set of variables involved for the plant and network shown in Figure 5.17 is as follows, the I.D. numbers being those assigned to the various nodes:

I.D.	Variable Description	Units
1.	Fuel Gas Heat Flow to gas turbine	MBTU/h
2.	GTG Power	MW
3.	Nox Steam Flow	Klb/h
4.	GTG Nox Concentration	ppm
5.	GTG particulates	ppm
6.	HP Steam Flow	Klb/h
7.	LP Steam Flow	Klb/h
8.	STG Throttle Flow	Klb/h
9.	Nox Extraction Flow	Klb/h
10.	Process Steam Extraction Flow	Klb/h
11.	Induction Steam Flow	Klb/h
12.	Condensate Flow	Klb/h
13.	STG Power	MW
14.	Nox Steam PRV Flow	Klb/h
15.	Dump Flow	Klb/h
16.	Process Steam PRV Flow	Klb/h
17.	Auxiliary Power Consumption	MW
18.	Total Feed Heater Steam Flow	Klb/h
19.	Power Exported to Utility	MW
20.	Process Steam Export	Klb/h

This set of variables is assigned to the columns of the matrix. The set of 27 equations, some of which are equalities and some inequalities, which define the network shown in Figure 5.17 is as follows:

Y.1 Linearized relationship between gas heat input and GTG power

Y.2 Relationship between gas heat and the amount of Nox steam admitted to the combustor

Y.3 Upper limit on Nox steam flow

Y.4 Auxiliary power consumption (could be a function of total power)

Y.5 Upper limit on GTG power

Y.6 Lower limit on GTG power

Y.7 Relationship between Nox emissions from GTG and gas heat input

Y.8 Relationship between particulate emissions from GTG and gas heat input

Y.9 Linearized relationship between H.P. steam generated and GTG heat input

Y.10 Linearized relationship between L.P. steam generated and GTG heat input

Y.11 Heater steam consumption as a function of amount of HP and LP steam generated

Y.12 Steam turbogenerator characteristic curve

Y.13 Steam turbogenerator steam/condensate balance

Y.14 Upper limit on Throttle steam flow

Y.15 Upper limit on Nox steam extraction

Y.16 Upper limit on process steam extraction

Y.17 Upper limit on exhaust steam flow to the condenser

Y.18 Lower limit on exhaust steam flow to the condenser

Y.19 Upper limit on steam turbogenerator power

Y.20 Lower limit on steam turbogenerator power

Y.21 Process steam header balance, including desuperheating water flow

Y.22 LP steam header balance

Y.23 HP steam header balance

Y.24 Nox steam header balance

Y.25 Power balance

Y.26 Export power target

Y.27 Export process steam target

As already stated, the only variables assigned a cost in the cost function are those variables which lie at the boundary of the problem, namely:

Fuel gas heat flow to gas turbine   +$/MMBTU/h

Tie line energy revenue $/mwh   -$/MWh
Revenue from export of process steam -$/klb/h

Note that while costs are assigned positive values, revenues have to be assigned negative values.

5.4.4 Results

Section 5.4.1 shows the complete set of results produced by a program structured in this way. Before the first iteration, the LP matrix is initialized with updatable constants which give an initial result, the cost being shown as -$921.05. The value of 30.19 MW for gas turbogenerator load contained in the solution is then used as the basis for developing the updated constants for the second iteration, the new equations being shown at the bottom of the section headed "Second Iteration". The cost after the second iteration is shown to be -$848.57.

The new gas turbogenerator load is 32.516 MW and this is used to update the coefficients once again. The cost after the third iteration is $848.12. The process is then repeated for a fourth iteration, the cost for which is again $848.12. Since the last two costs were sensibly the same, the program terminates, the final set results at convergence being shown in Table 5.4.1.

Table 5.4.2 shows the results when the plant is running at full load but under a variation of export process steam flows and ambient temperatures. Column A shows how the plant is loaded when generating 50 MW and delivering zero process steam flow. Most of the Nox steam is extracted from the STG, the balance being made up by a small flow through the PRV on node #14, plus water passed to the desuperheater to maintain Nox steam temperature.

Column B shows that, when generating 50 MW and delivering 30 Klb/h of process steam, the Nox steam can be supplied more economically by desuperheating live steam passed

Table 5.4.1   Results summary

Ambient temperature	70.0 Deg.F
Cost is	−848.12
Fuel Gas Flow	375.474
GTG Power	32.520
Nox Steam Flow	23.545
GTG Nox	10.852
GTG particulates	5.973
HP Steam Flow	139.444
LP Steam Flow	22.726
STG Throttle Flow	119.432
Nox Extr. Flow	.000
Process Extr.	86.967
Injection Flow	13.972
Condensate Flow	119.437
STG Power	8.480
HP Press. RV Flow	20.042
Dump Flow	0.000
Process Stm PRV	0.000
Auxiliary Power	1.000
Heater Steam	8.764
Utility Export MW	40.000
Process Steam Export	100.000

through the PRV at node #14. In fact all the remaining results show this to be the preferred method of supplying Nox steam and the cost of this port could have been eliminated from the original design. Furthermore, the Nox steam extraction causes a much more severe reduction in power generation for a given throttle flow than does the process steam extraction; so that backing off the Nox extraction increases the amount of power which can be generated. Finally, the use of high pressure steam with a high percentage of desuperheating water being added is often a very cost effective way of supplying steam for process needs, in this case the Nox steam.

Table 5.4.2   Full load and varying process steam flows

Description	A	B	C	D
Total Export MW	50.	50.	50.	50.
Total Export Process Steam	0.	30.	40.	40.
Cost	-884.38	-986.66	-1012.0	-1018.8
Ambient temperature	70.	70.	70.	80.
Fuel gas flow	403.9	404.58	407.0	405.30
GTG Power	36.1	36.1	36.39	36.10
Nox steam flow	26.81	26.885	27.16	26.93
GTG Nox	11.67	11.69	11.76	11.71
GTG Particulates	6.42	6.43	6.47	6.44
HP Steam Flow	150.89	150.84	151.77	155.99
LP Steam Flow	22.91	23.68	23.81	23.42
STG Throttle flow	148.85	124.84	128.69	127.76
Nox Extraction flow	22.01	0.	0.	0.
Process extraction	0.	26.09	34.78	29.43
Injection flow	13.52	14.25	14.32	13.74
Condensate flow	140.36	113.01	108.24	112.06
Power	14.9	14.9	14.61	14.9
HP Pressure Reducing Flow	2.04	22.84	23.08	22.88
Dump Flow	0.	3.16	0.	0.
Process Steam PRV	0.	0.	0.	5.34
Heater steam	9.38	9.42	9.48	9.69
Auxiliary Power	1.0	1.0	1.0	1.0

It should also be noted that in the optimum and least cost solution shown in Column B there is a small flow through the dump valve of node #15. This is because the turbogenerators are constrained towards their upper limits while the quantity of high pressure steam generated in the HRSG has not changed. Water is also being added to the process steam extraction flow to maintain its temperature; so that, since it can not be passed to the condenser to generate more power, the STG already being at its power limit, there is a balance of high pressure steam that has to be dumped.

Table 5.4.3   Part load and varying process steam flows

Description	E	F	G	H
Total Export MW	40.	40.	30.	30.
Total Export Process Steam	100.0	100.0	100.	100.
Cost	-848.12	-857.02	-586.65	-598.10
Ambient temperature	70.	80.	70.	90.
Fuel gas flow	375.47	373.25	315.84	312.98
GTG Power	32.52	32.1	24.81	23.97
Nox steam flow	23.54	23.23	16.76	16.24
GTG Nox	10.85	10.79	9.13	9.05
GTG Particulates	5.97	5.94	5.02	4.98
HP Steam Flow	139.44	143.3	115.62	121.66
LP Steam Flow	22.72	21.94	22.00	22.28
STG Throttle flow	119.43	123.56	101.38	107.92
Nox Extraction flow	0.	0.	0.	0.
Process extraction	86.96	86.96	86.96	86.96
Injection flow	13.97	13.02	14.57	14.46
Condensate flow	46.44	49.62	28.99	35.36
Power	8.48	8.89	6.19	7.03
HP Pressure Reducing Flow	20.0	19.74	14.24	13.79
Dump Flow	0.	0.	0.	0.
Process Steam PRV	0.	0.	0.	0.
Heater steam	8.76	9.92	7.43	7.77
Auxiliary Power	1.0	1.0	1.0	1.0

As shown in Column C, when the export process steam flow is increased to 40 Klb/h this dumping ceases. There is a also slight redistribution of power generation between the two machines since the increased extraction flow reduces the capacity of the STG.

In column D, data is taken for the same case as column C but with the ambient temperature rising to 80 Deg.F. It should be noted that the revenue has increased slightly from $1012/h to $1019/h and that the amount of high pressure steam generated in the HRSG has also increased. In other words, the heat recovery of the cycle has improved.

Table 5.4.3 shows the results with the plant running under partial load and with 100 Klb/h of process steam. Column E is the same case as was included in Section 5.5.1. Column B again shows the effect of increasing ambient temperature from 70 to 80 Deg.F, resulting in the revenue also increasing from $848/h to $857/h.

In column G the amount of exported power has been reduced to 30 MW. Column H is the same case as Column G but the ambient temperature has been raised from 70 to 90 Deg.F, the revenue also increasing from $586.6/h to $598.1/h, an increase of $11.5/h. It is suggested that a case can be made for recycling some of the HRSG exhaust gases to the compressor inlet in order to improve the cycle heat rate when operating under partial load conditions.

Other evaluations of the program have drawn attention to when it is NOT possible to satisfy a desired combination of export power and process steam.

5.4.4.1   On-line Application

The most comprehensive on-line application of the above technique was provided to The Midland Cogeneration Venture and was reported by Allen and Polito, et al (1992). This included automatic updating of the gas turbogenerator characteristic curves, allowing the solution to respond to changes in turbogenerator heat rate caused by fouling. This is an important feature which, if not performed, can result in suboptimal solutions. In all on-line applications an off-line version of the program is available so that management and engineers can evaluate how future combined energy demands can be met, even when some of the equipment is out of service.

Many plants are provided with more than one train of equipment. A convenient method of excluding an equipment item from the solution is to create one dummy variable for each equipment item which might need to be excluded and assign it a value of zero or unity. If all of the constraints and intercepts for a unit are marshalled in the appropriate column, then setting the dummy variable to unity includes the unit while setting it to zero excludes the unit completely. (See SW of Table 4.2)

5.4.4.2   Accuracy of the Method

The accuracy of the method is not dependent on the linear programming algorithm but on the detailed design of the matrix, on the accuracy of the models and also on the costs assigned to the appropriate resources. The matrix must contain all of the pertinent variables together with a complete set of the upper and lower constraints that must be observed, otherwise the optimal solution can not be implemented in practice. All of the mass, power and heat balances must also be included, although the heat balance equations within the matrix are implied as, for example, within the steam turbogenerator characteristic curve.

The importance of model accuracy cannot be too strongly stressed. Signal validation and data reconstruction techniques should be implemented to verify the accuracy of the data inputs, used by the regression analysis program to periodically update equipment models that might have changed, for example, due to fouling. Inaccurate models can cause a sub-optimal set of assignments to be generated even with a successful convergence of the program.

Extensive experience of using the LP algorithm to optimize energy processes has shown that the relative improvement in cycle performance can be relied upon with confidence. It has also been found that the magnitude of the savings estimates tends to be conservative since certain beneficial side effects resulting from optimal operation are not expressly reflected in the equation set and therefore in the estimates.

5.4.5 Optimization Results

Objectives:

Total Power: 40.0 MW
Exported Process Steam Flow: 100.0 Klb/h
Enter Ambient Temperature: 70.0 Deg.F

First Iteration:

Initial Cost: -$921.05

Second Iteration:

GTG Power = 30.192 MW
Exhaust Flow = 1295.975 Klb/h Maximum Power = 42.804 MW
Exhaust Temp. = 838.074 Deg.F   Heat Rate = 11826.250 BTU/kwh
Nox ratio = 1.27506   Nox Flow = 21433.25 lb/h
HP steam flow = 132.232 Klb/h LP steam flow = 22.258 Klb/h

GTG Power = 31.192 MW
Exhaust Flow = 1295.975 Klb/h Maximum Power = 42.804 MW
Exhaust Temp. = 850.784 Deg.F
HP steam flow = 135.294 Klb/h LP steam flow = 22.492 Klb/h

GTG Equations:
  Fuel (MBTU/h) - 7.874908 * MW   = 119.29990
  .1145976 * Fuel (MBTU/h) - Steam Flow (Klb/h) = 19.48506
  Maximum Power   = 42.80461

HRSG Equations:

H.P. Evaporator Equation: HP STEAM FLOW (Klb/h) - 3.06172*
  MW = 39.79218
L.P. Evaporator Equation: LP STEAM FLOW (Klb/h) - .23288*
  MW = 15.22745

Cost: -$848.57

Third Iteration:

GTG Power = 32.516 MW
Exhaust Flow = 1295.975 Klb/h Maximum Power = 42.804 MW
Exhaust Temp = 838.074 Deg.F Heat Rate = 11547.170 BTU/kwh
Nox ratio = 1.331   Nox Flow = 23544.04 lb/h
HP steam flow = 139.435 Klb/h LP steam flow = 22.723 Klb/h
GTG Power = 33.516 MW
Exhaust Flow = 1295.975 Klb/h Maximum Power = 42.804 MW
Exhaust Temp. = 880.783 Deg.F
HP steam flow = 142.639 Klb/h LP steam flow = 22.838 Klb/h

GTG Equations:

   Fuel (MBTU/h) - 8.035248 * MW = 114.19250
   .1148400 * Fuel (MBTU/h) - Steam Flow (Klb/h) = 19.57423
   Maximum Power = 42.80461

HRSG Equations:

H.P. Evaporator Equation: HP STEAM FLOW (Klb/h) - 3.20465*
   MW = 35.23325
L.P. Evaporator Equation: LP STEAM FLOW (Klb/h) - .11495*
   MW = 18.98512

Cost is: -$848.12

Fourth Iteration - Convergence and Solution:

GTG Power = 32.516 MW
Exhaust Flow = 1295.975 Klb/h Maximum Power = 42.804 MW
Exhaust Temp. = 867.800 Deg.F Heat Rate = 11547.090 BTU/kwh
Nox ratio = 1.331 Nox Flow = 23544.69 lb/h
HP steam flow = 139.437 Klb/h LP steam flow = 22.272 Klb/h

GTG Power = 33.516 MW
Exhaust Flow = 1295.975 Klb/h Maximum Power = 42.804 MW

Exhaust Temp. = 880.792 Deg.F
HP steam flow = 142.641 Klb/h LP steam flow = 22.838 Klb/h

GTG Equations:

Fuel (MBTU/h) - 8.035309 * MW   = 114.19040
.1148403 * Fuel (MBTU/h) - Steam Flow (Klb/h) = 19.57436
Maximum Power   = 42.80461

HRSG Equations:

H.P. Evaporator Equation: HP STEAM FLOW (Klb/h) - 3.20468*
MW = 35.23217
L.P. Evaporator Equation: LP STEAM FLOW (Klb/h) - .11495*
MW = 18.98515

Solution Cost is:-$848.12

5.5 Condenser/Cooling Tower Subsystem Optimization

In the typical condenser/cooling tower subsystem there are often several cooling water circulation pumps operating in parallel with one another; while the cooling tower often has several fans or one fan able to operate at several speeds. During the summer time the amount of heat to be removed to condense the exhaust from the steam turbine often requires that all circulating water (c.w.) pumps are running and that all fans be in operation and, where appropriate, running at full speed. However, under winter conditions it will often be found possible to remove the same amount of heat from the exhaust with only one c.w. pump running; while some of the cooling tower fans can also be switched off. Depending on the circumstances at a particular site, the savings in auxiliary power costs can be significant if it is possible for some auxiliary equipment motors to be shut down.

5.5.1.1 Condenser Model

Starting with all c.w. pumps running, a Newton-Raphson model of the condenser can be written in which the principal boundary conditions are:

- Amount of heat to be removed from the steam turbine exhaust
- Cooling water flow rate
- Desired condenser back pressure and corresponding vapor temperature

The basic heat equations that have to be satisfied are those in (3.23) and (3.24). Assuming that all c.w. pumps are running, the model will calculate the corresponding inlet and outlet c.w. temperatures for this condenser (T_{in} and T_{out} respectively) when running under the stated set of boundary conditions. These then become the outlet and inlet water temperatures that the cooling tower has to satisfy.

5.5.1.2 Cooling Tower Model

The cooling tower characteristic curve will be known (equation (3.35)). Starting with all fans running, the corresponding air flow (G) will be known while the water flow rate (L) will be that equivalent to the number of c.w. pumps operating. Thus the ratio (L/G) will be known so that the value of KaV/L can be calculated from equation (3.35). Assuming that the c.w. outlet temperature from the condenser (T_{out}) is the water inlet temperature to the cooling tower, it is possible to use equation (3.34) to calculate the temperature of the water leaving the cooling tower that will satisfy the value of KaV/L.

If this temperature is less than T_{in} then some overcooling would occur if all fans were running. The cooling tower model can then be re-executed with a fewer number of fans (and corresponding value of (G)), the number of fans being decreased until the calculated cooling tower outlet temperature is higher than T_{in}. Thus the minimum number of fans needed to satisfy the conditions with all c.w. pumps running can be determined.

The number of c.w. pumps may then be reduced (together with the corresponding value of L) and new values of T_{in} and T_{out} calculated using the condenser model. These can then be presented to the cooling tower model to calculate the minimum number of fans that will satisfy the conditions when fewer c.w. pumps are running.

The optimum combination of auxiliaries is the one that will remove the required amount of heat from the turbine exhaust at the least cost in terms of the amount of auxiliary power consumed.

5.5.1.3   *Commentary*

In any given plant there may be more optimizing opportunities than the above suggests. For instance, Figure 3.1 includes a feature that allows the c.w. inlet temperature to the condenser to be controlled at some value higher than the cooling tower outlet temperature, by allowing a portion of the water discharged from the condenser to be mixed with the water leaving the cooling tower. This feature may be useful under some winter conditions to prevent the water inlet temperature to the condenser from becoming so low that under-cooling of the condensate might occur. Under-cooling of the condensate is wasteful since the excess heat removed must be replaced in the boiler.

Another possibility would be to add the minimum c.w. inlet temperature as a boundary condition to the condenser model, the inlet temperature being greater than this boundary condition on model convergence. In either case, there will be a tendency for fewer fans to be in operation since less heat has to be removed by the cooling tower.

Chapter 6
Controlling the Steam
Turbogenerator for Watts, VARS,
Volts and Frequency

6.1 Introduction

The following is intended as a brief introduction to the operation of power generation and distribution systems within industrial plants, particularly those with steam turbogenerators. It is especially intended to provide a framework for understanding the principles involved in adjusting turbogenerator excitation so as to regulate bus voltage and/or machine power factor; and to indicate the various constraints on excitation adjustment. From this understanding are defined the principles for using excitation to optimize plant and tie-line power factor in industrial plants. Note that the words "alternator" and "generator" are used synonymously in the following text.

As a brief historical note, Thomas Alva Edison (1846–1931) invented his earliest generator around 1879, as a means for supplying power to the incandescent lights he had recently invented. This generator, an example of which is shown in Figure 6.1, was of the direct current two-wire type and generated power at 240 vDC. The addition of a balance coil enabled a 3-wire system to be provided, operating at 120 vDC.

In his Edison biography, Baldwin (1995) introduces a contemporary of Edison's, Nicola Tesla (1856–1943), who

Figure 6.1   Early Edison Generator - ca. 1890, courtesy –
Edison-Ford Winter Estates, Fort Myers, FL

was born in Croatia and studied at the Austrian Polytechnic School in Graz. His vision was of a polyphase alternating current system and, he joined Edison in New York but Edison continued to favor the direct current principle and was hostile to Tesla's approach. A modern version of Tesla's generator, designed to generate alternating current, is shown in Figure 6.2. Tesla's breakthrough came in 1888 when he delivered a lecture to the American Institute of Electrical Engineers on "A New

Figure 6.2   Modern Turbogenerator, courtesy of Lakeland Electric Company

System of Alternate Current Motors and Transformers". Tesla's lasting contribution to engineering in the United States was the hydroelectric power generation and transmission system located in Niagara Falls, NY.

Meanwhile, George Westinghouse (1846–1914) in 1876 started to build up the Westinghouse Electric Corporation based on an alternating current generation and transmission system. Tesla's lecture caught his attention and he hired Tesla as a consultant for a fee of $2000-per-month, also purchasing the manufacturing rights to Tesla's single and polyphase motors and transformer designs. By 1889, Westinghouse had installed 1,100 new electric power plants across the country and the rest is history.

6.2   Turboalternators

The alternators driven by steam turbines consist of a stator that is fixed and a turbine-driven rotor, consisting of windings mounted on a shaft. The windings of the rotor are supplied with an adjustable direct current for excitation (See Figure 6.3), this current commonly being provided by a small DC excitation generator mounted on or coupled to the turbine shaft. The stator also contains windings typically arranged in four poles and three phases. The windings on the rotor are typically arranged as a single pole and as the magnetic field created by the excitation current rotates at 1800 rpm inside the stator, it causes three-phase voltages and currents to be generated in the stator windings at 60 Hz.

The *load* currents generated in the stator set up a field which exerts a magnetic torque on the shaft, this being the torque which is being provided by the turbine. The *speed* of the machine is either controlled by tie-line *frequency*; or the

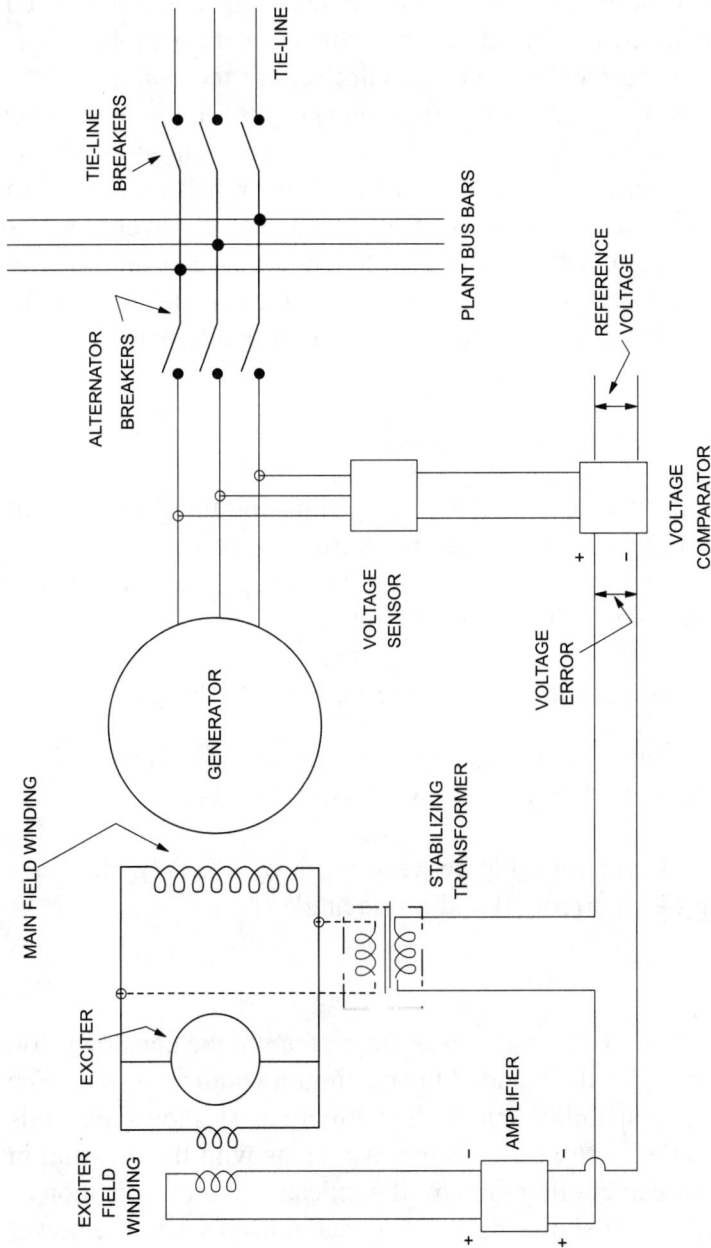

Figure 6.3   Typical generator/exciter configuration

governor on the machine is used to determine frequency when the machine is islanded and not connected to a tie-line. The excitation current is used to control either the *voltage* at the terminals of the alternator (generator) or else the *power factor* of the generated load. In some cases the excitation current is used to control both as described below. When connected to the tie-line, it will be shown that the alternator power factor is usually different from both the plant power factor and the tie-line power factor, and is one of the means that is used to regulate the latter. To summarize, a turboalternator is controlled so as to:

- When connected to a tie-line:

  o Regulate generated power by adjusting the governor that supplies steam to the turbine throttle valve
  o Since the tie-line determines the voltage, use excitation to regulate tie-line power factor

- When disconnected from the tie-line:

  o Regulate plant frequency by adjusting turbine speed
  o Regulate bus voltage by adjusting excitation

The basic relationship between megawatts (MW), alternator voltage (V), current (I) and phase angle (ϕ) is:

$$MW = V * I * \cos\phi \tag{6.1}$$

where $\cos(\phi)$ is termed *power factor*. One of the constraints on an alternator is the value of the maximum winding current. The reactance capability curve (See Figure 6.4) shows that this value varies with power factor, as well as with the pressure in the hydrogen cooling system, if applicable. An examination of Equation (6.1) shows that MW is maximized when the power factor approaches unity.

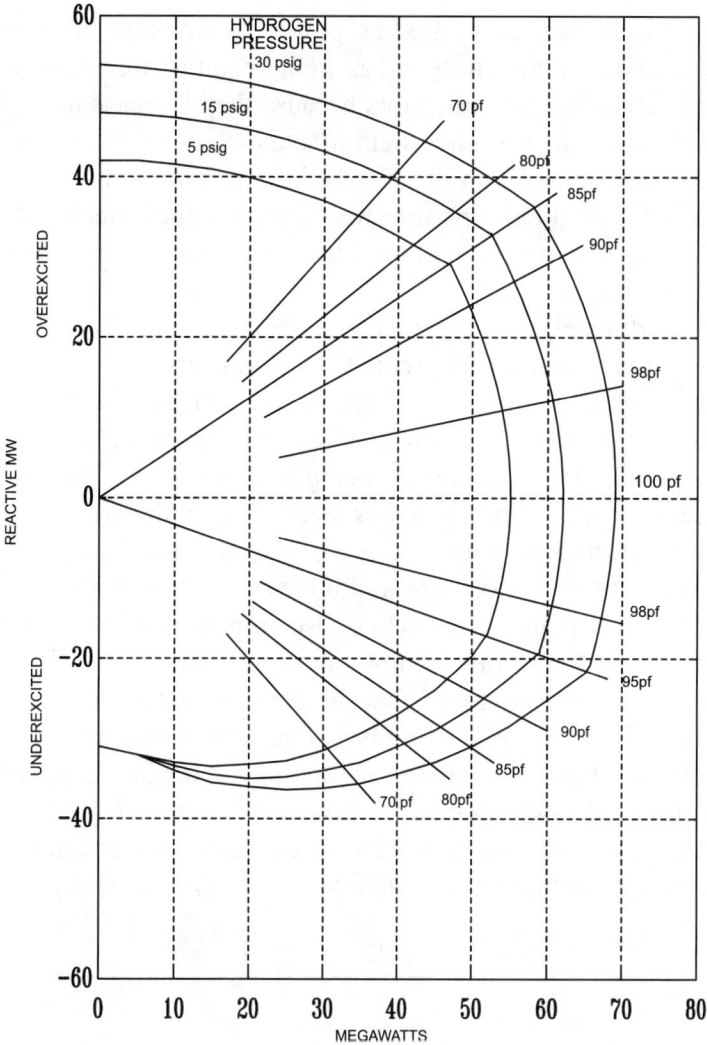

Figure 6.4   Alternator reactance capability curve

Thus, an increase in alternator power factor not only allows the generated load to be maximized for the maximum allowable current in the stator windings (See Figure 6.4), but also reduces the electrical losses in the alternator for a given load.

Similarly, increasing tie-line power factor often has a beneficial effect on the utility bill as well. Penalties are customarily imposed for low power factors because they increase the utility transmission losses and their associated costs. Low power factors also cause a utility to provide transformers ad switchgear with a higher rating than necessary. On the other hand, the utility will often offer credits if the plant contracts to maintain high power factors.

Unfortunately, when generators are being used to control tie-line power factor, the reduction in purchased VARS, or reactive power, must be offset by an increase in generated VARS. Thus the power factor within the alternator is reduced while that of the tie-line is being increased, resulting in an increase in the winding losses even though the transmission losses incurred by the utility are being reduced. This suggests that one of the economic tasks for the power house operators is to minimize plant losses while also minimizing the penalties imposed by the utility for low power factors.

It has to be recognized that the control strategy for a given plant will depend on the number of generators running on-line, the bus configuration or system topology and whether the plant is connected to the tie-line, if provided. Sections 6.5 take note of the considerations that have to be taken into account in a variety of configuration situations.

6.3   Starting Up a Generator

To bring an industrial power plant on line from a cold start requires that steam be raised in the boilers, although this itself requires power to be available to run the boiler auxiliaries such as feedwater pumps, fans and mechanical handling equipment. No difficulty is encountered when a tie-line is available; but

in those cases with no available external source of power, diesel-
or gas-turbine driven generators must be installed. Steam
turbogenerator-driven ships, mining installations in the wilder-
ness and the occasional paper mill designed to be energy self-
sufficient are examples of the latter. However, whether a tie-line
or auxiliary generators are present, each steam turbogenerator
being brought on line must always first be synchronized with the
plant busses before the generator can be connected to the plant
electrical distribution system and allowed to carry load.

6.4   Synchronizing Process

When bringing a steam turbogenerator on-line, it must first be
warmed up with the turbine auxiliary equipment running, the
shaft being rotated under the action of the turning gear. Once the
specified steam/metal temperature differences and shaft expan-
sions have been attained, the throttle valve can be carefully
opened and the machine brought up to speed. However, before
the breakers can be closed to connect the alternator to the
distribution system, three electrical conditions have to be met:

- The speed must be such that the machine frequency is the
  same as that at the bus bars
- The voltage at the alternator terminals must be equal to that
  on the bus bars
- The breaker must only be closed at or very near the instant
  when the generated voltage and the bus bar voltage are in a
  close phase relationship with one another

The speed and, therefore, frequency can be regulated by
adjusting the setting of the governor; while the machine voltage
can be regulated by adjusting the excitation. However, Hirst
(1942) shows that achieving the correct phase relationship is

more difficult and requires the presence of synchronizing lamps (See Figure 6.5) or a synchroscope (See Figure 6.6).

In the case of the older systems consisting of the three lamps shown in Figure 6.5, the upper lamp is connected directly across the breaker on phase B; while the other two are across phases A and C of the breakers. If the generator being connected is running at a higher frequency than the distribution system, the lamps will appear to be rotating counter-clockwise and at a speed corresponding to the difference between the machine and system frequencies. Conversely, if the generator being connected is running slow, the lamps will appear to rotate clockwise. The correct moment for synchronizing is when the lamps appear to be stationary *and* the lamp directly connected across phase B is dark.

In the case of the more modern synchroscope (Figure 6.6), when the alternator and system voltages are in phase, Coil B exercises torque while coil A, being in quadrature with the stator flux, exercises little or no torque. The result is that the rotor moves until coil B is in line with the stator field and the pointer is vertical. Differences in phase cause the rotor to rotate, again at a speed corresponding to the frequency difference. When synchronizing, the speed of the incoming machine is adjusted until the rotation of the pointer is very slow, the breaker being closed only when the pointer is actually passing through the vertical position.

6.5   Single Steam Turbogenerator with No Tie-Line

As soon as the incoming machine is connected to the busses a small amount of current will start to be generated. The load can now be increased by adjusting the governor so as to slightly

NOTE: AS B-B' GOES TO ZERO
INDICATING VOLTAGES ARE IN PHASE
B-LIGHT GOES DARK

MAIN BUS BARS

SYNCHROSCOPE

MAIN
BREAKERS

A' B' C'

A B C

GENERATOR

(a)

(b)

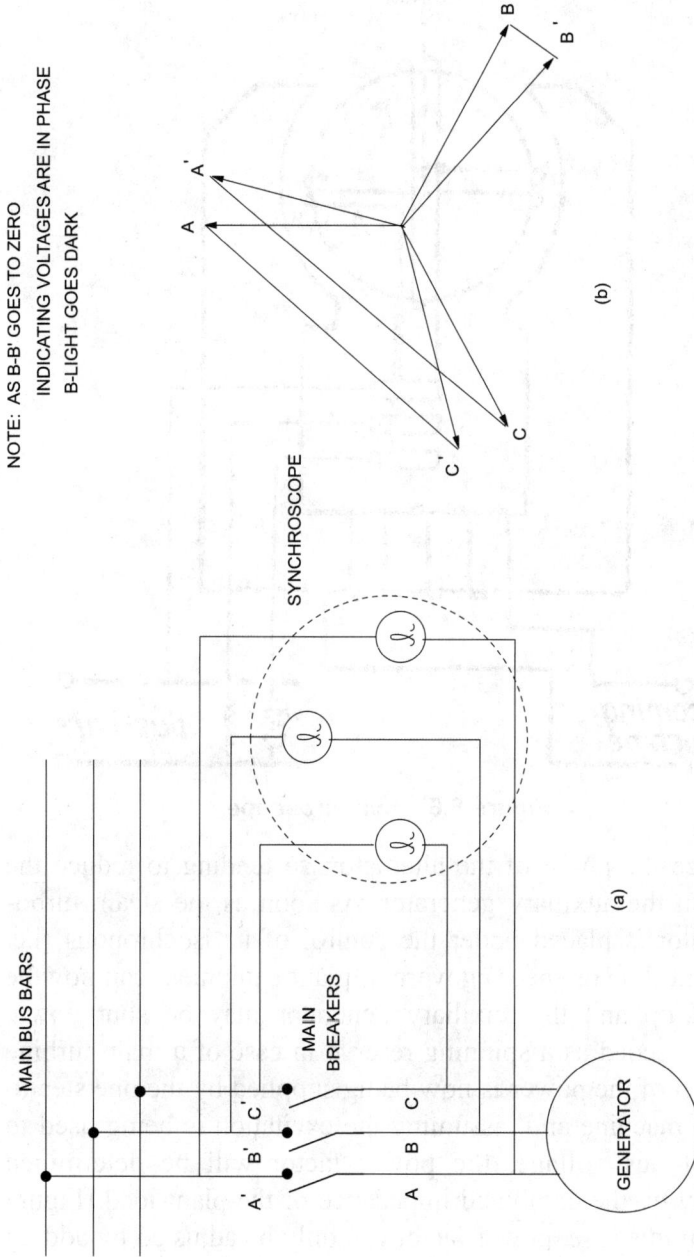

Figure 6.5 Synchronizing of three-phase alternator using lights

Pointer

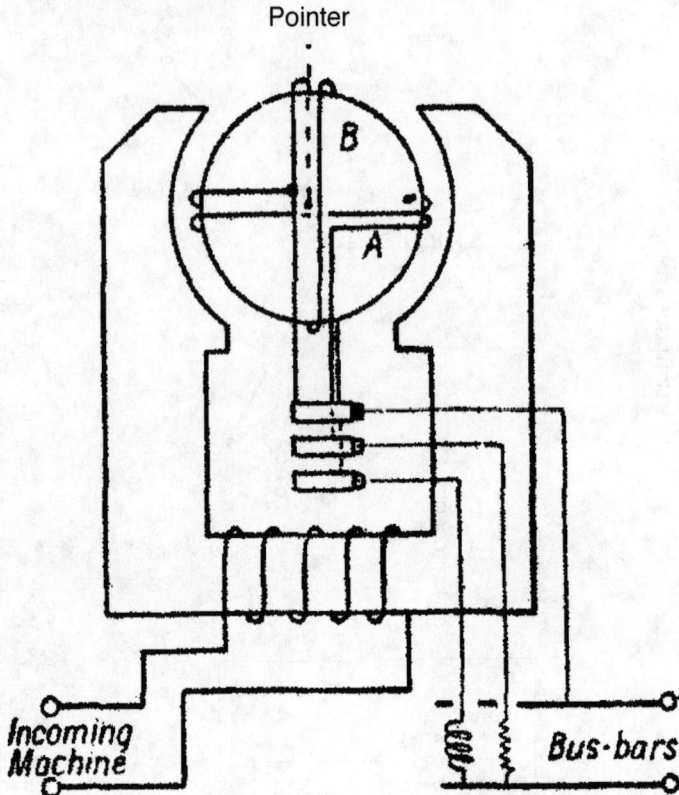

Figure 6.6   Synchroscope

advance the phase of the alternator, so tending to reduce the load on the auxiliary generator. As soon as the steam turbo-generator is placed under the control of its isochronous (i.e. constant time or speed) governor, plant equipment can now be started up and the auxiliary generator may be shut down, unless needed as a spinning reserve in case of a main turbine trip. All of the power is now being supplied by the one steam-driven machine and, assuming the excitation is being used to control bus voltage, the power factor will be determined entirely by the combined impedance of the plant load (Figure 6.8). In this case, power factor can only be adjusted by adding capacitor banks or by means of some other device such as a

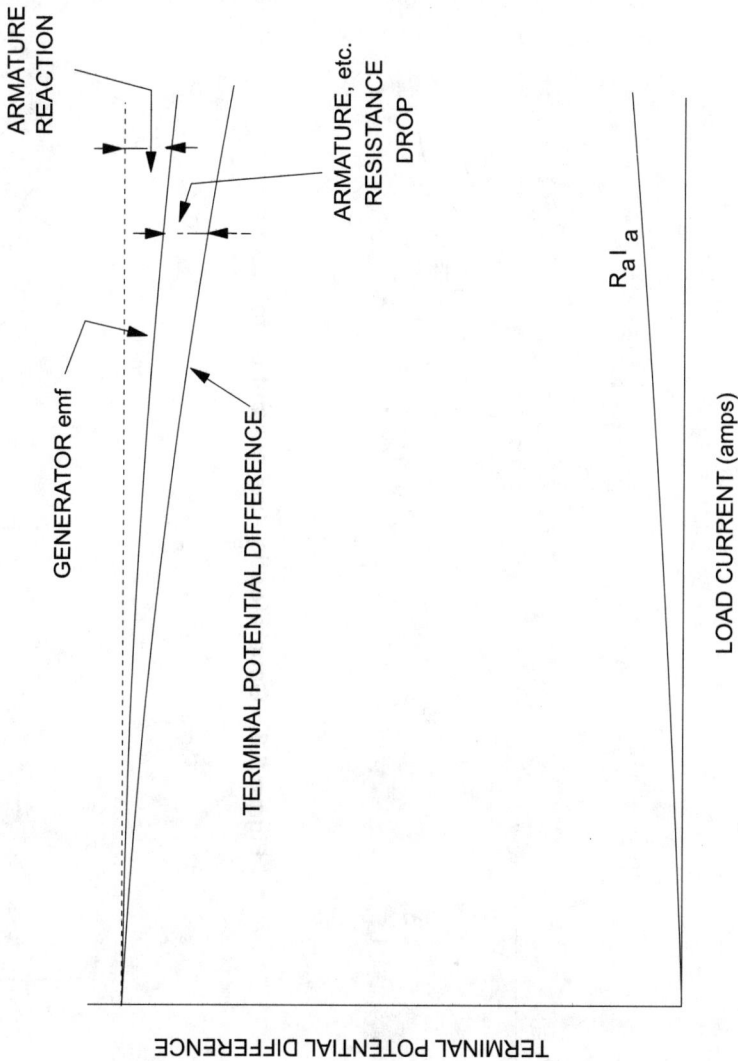

Figure 6.7   Load characteristic of separately excited generator

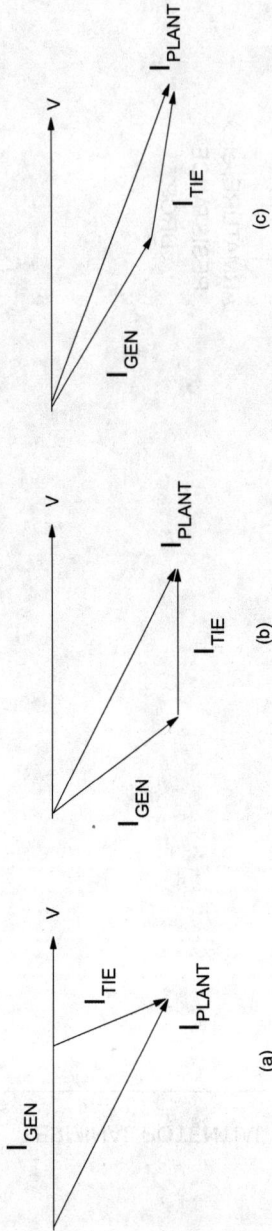

Figure 6.8 Generator/tie-line current distributions

synchronous condenser, i.e. by adjusting the total impedance of the plant load. Thus, in this plant configuration:

- The governor controls speed and hence frequency
- The excitation controls bus voltage, subject to maximum excitation current, and
- The generator and system power factor can only be controlled by adjusting plant load impedance, and not by adjusting generator excitation.

6.6   Single Generator in Parallel with Tie-Line

As in the above case, as soon as the incoming machine is connected to the distribution busses, a small amount of current will start to be generated. Again, the load can be increased by adjusting the governor to slightly advance the phase of the machine and this will tend to take the load off the tie-line. More plant equipment can now be started, the immediate effect of this increase in load being that more power is drawn from the tie-line. This can be reduced by adjusting the governor of the machine so that the latter picks up load; but the drooping characteristic of an alternator (See Figure 6.7) may mean that the excitation also needs adjustment.

Note that with the tie-line connected, the tie-line alone controls the frequency and, whether the alternator lags or leads the tie-line, a synchronizing current is automatically set up within the alternator. It is this synchronizing current that causes each generator to remain synchronized with the tie-line, unless over-speeding or some other governor fault should develop.

Since the tie-line can be considered an infinite bus, it also essentially determines the bus and generator voltages. Thus, prior to synchronization, although the excitation was used to adjust the machine voltage: once coupled to the tie-line the excitation may be used to adjust the power factor of the machine and hence of the tie-line. The phasor diagram of Figure 6.8 explains the principle. In Figure 6.8a, the generator is adjusted to unity power factor with the current in phase with the bus voltage, all the VARS required by the plant load being supplied by the tie-line. In Figure 6.8b, the power factor of the machine is adjusted so that all of the VARS are being supplied by the generator, the power factor on the tie-line rising towards unity. Normal practice is to adjust the excitation to somewhere between these two extremes, as shown in Figure 6.8c.

While one of the considerations in excitation control is the choice between importing or generating the VARS needed by the plant, another concerns the maximum current capacity of the generator windings, sometimes referred to as the reactance capability (Figure 6.4). As already mentioned, the effect of this is that as the load on the generator increases, its reactance capability decreases and becomes a constraint when the power factor is less than unity. In this case, it may no longer be possible to hold the tie-line power factor at its desired value. It should be borne in mind that another important constraint is that of maximum excitation current.

In the event that the tie-line breaker trips, the generator should be immediately switched to isochronous control. It will now try to pick up the load in its effort to maintain frequency and the excitation may need to be adjusted to maintain bus voltage. However, should the available generating capacity be insufficient to support the load and the frequency drops, plant loads will now have to be shed to bring the internal power system back into equilibrium. Once this has been achieved, control of excitation will once more revert to that of maintaining the bus voltage, as in the case 6.5 above where no tie-line was connected.

It should be noted in passing that the reactance capability of hydrogen-cooled generators varies with the hydrogen pressure, as shown in Figure 6.4. This should be taken into account when determining the upper reactance constraint.

6.7 Several Machines Operating in Parallel with No Tie-Line

Where there is no tie-line present and more than one machine is to be running in parallel, only one of them is used to control the frequency the other(s) being held in synchronism with the frequency-controlled machine. The synchronizing procedure outlined in Section 6.4 above must of course be used to connect each incoming machine when it is brought on line.

As shown in Figure 6.7, machines not equipped with an isochronous governor exhibit a drooping characteristic, the speed falling slightly with an increase in resistive load. Thus kilowatt load distribution among the machines can be controlled by adjusting the set points of the governors on all except for the frequency-controlled machine, in accordance with the results calculated by an economic dispatch program. Note that the load on the frequency-controlled machine will be the difference between the total plant load and the loads being generated by the other machines now running on-line.

The excitation is adjusted to maintain the voltage on the bus of each machine at the required value, subject again to the constraint due to maximum excitation current. The power factor on each machine is dependent on the total impedance of the plant load as well as being a function of bus voltage and, hence, machine excitation. In summary, in this configuration:

- Frequency is controlled by only one machine
- The voltage on the bus of each machine is controlled by adjusting the excitation of that machine

- The loads on other than the frequency-controlled machine can be regulated by adjusting their governor setpoints
- The power factor on each machine is a function of voltage, hence excitation, and also dependent on plant system impedance

---

6.8 Several Machines Operating in Parallel with Tie-Line Connected

The operation in this case is similar to that in Section 6.6 except that none of the machines would be set to frequency control. Thus loads can be freely assigned among all machines operating in parallel and the governor settings will be adjusted in accordance with the results from an economic dispatch program.

As in the case of Section 6.6 with one machine operating in parallel with the tie-line, tie-line power factor can be controlled by means of generator excitation. Again, care must be taken to observe the reactance capability limits of each machine as well as its maximum excitation current.

6.9 Conclusion

1. Where one machine is generating power in a plant which is not connected to a tie-line, power factor can not be controlled by adjusting excitation.
2. Where one machine is generating in parallel with a tie-line, tie-line power factor can be controlled within the limits of machine reactance capability and maximum excitation current.

3. Where several machines are generating in parallel but without being connected to a tie-line, load distribution among them can be controlled by adjusting the governors on all but the frequency-controlled machine. Excitation is first used to control bus voltages subject to limitations on excitation current, the power factor on each machine being essentially dependent on plant system impedance.

4. Where several machines are generating in parallel with a tie-line, not only can the distribution of load be controlled but also tie-line power factor through adjustments to machine excitation. This must be done within the limits of reactance capability (i.e. maximum generator stator current) and excitation (i.e. rotor) current. Once the minimum tie-line power factor target has been determined, excitation should be increased on all machines, maintaining their power factors at almost the same values until either this target is reached or the maximum excitation current on any one machine is attained. Under steady state conditions, when the electrical constraints on all machines are being pushed, it is probable that safeguards will be needed to protect the alternators during load changes. For this reason an arbitrary low limit (e.g. 70%) should be placed on machine power factor. Otherwise, a feed-forward signal may be introduced into the machine control system so as to adjust excitation immediately in response to a load change, reducing excitation as the load increases.

5. In the case of hydrogen-cooled generators, the change in the reactance constraint as a function of hydrogen pressure should also be taken into account.

Chapter 7

The Importance of Process Heat Exchangers in Industrial Energy Systems

Heat exchangers are frequently included in the process flow sheets of industrial plants and have an important part to play. In cogeneration plants, the condensers attached to the exhaust from low pressure turbines recover not only the condensate, so reducing the cost of treating new makeup water, but also the heat that would otherwise be discharged into the environment and wasted. In many processes they are also used to minimize the heat lost to the atmosphere by recovering some of the heat contained in process discharge streams. Other heat exchangers are used to recover the heat contained in a process stream recycled from a late stage in the process and use it to preheat a stream entering an earlier stage in the process. In all cases, the increase in entropy engendered by the process is reduced and even minimized.

Heat exchangers can take many forms but, in all cases, the heat exchanger is presented with two fluid streams operating at different temperatures; the fluids are physically separated, the separation means also acting as a heat exchange surface. The separation commonly consists of bundles of tubes through which one of the fluids flows (See the shell and tube heat exchanger depicted in Figure 7.1). Sometimes finned tubes are used. The separation may also take the form of plates between which a fluid flows.

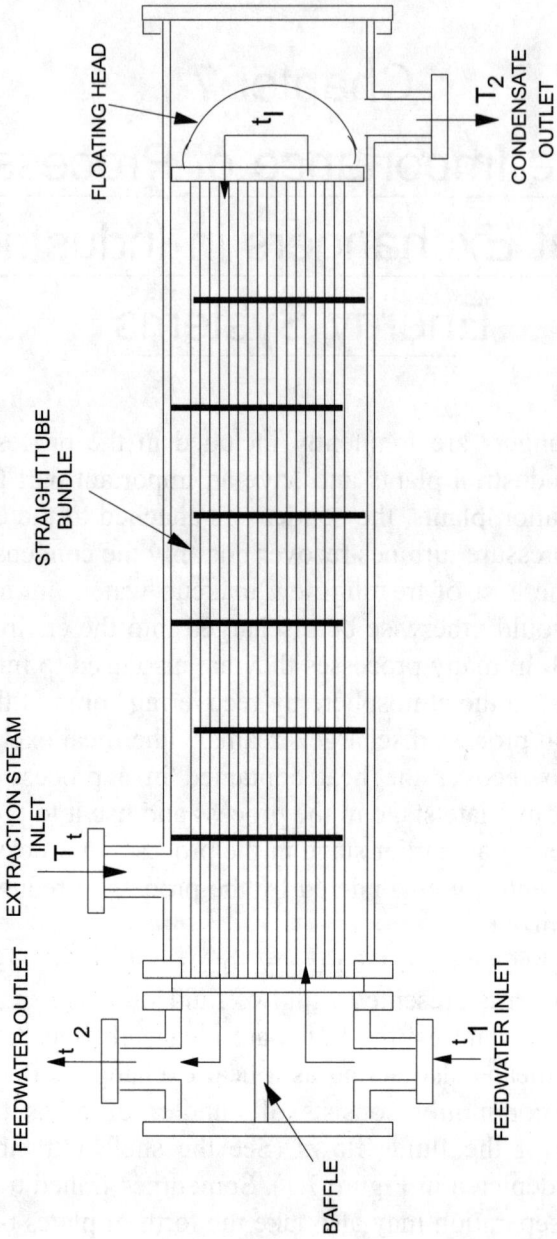

Figure 7.1   Double-pass horizontal feedwater heater straight tube bundle

7.1   Shell-and Tube Heat Exchangers

The form of the heat exchanger most commonly found in industrial processes is of the Shell and Tube type, shown in Figure 7.1, which is configured as a feedwater heater in a Rankine Cycle power plant, receiving steam extracted from the steam turbogenerator. The heat exchanger consists of a fixed head provided with inlet and outlet ports through which the cool fluid passes. The body or shell of the heat exchanger is also provided with inlet and outlet ports through which the heating fluid passes. Within the shell is a tube bundle, one end of each tube being attached to the fixed head while the other is often attached to a floating head that allows the tubes to expand and contract without damage. Figure 7.1 shows a bundle of straight tubes but the bundle can also consist of a set of U-tubes that are only attached to the fixed head. Clearly, where fouling is likely to occur, straight tubes are easier to clean than U-tubes but specially articulated cleaners have now been developed that allow the bends to be accommodated by the cleaners.

The shell is also provided with a set of baffles that not only act as tube supports but also cause the heating fluid to move through the tube bundle in a serpentine fashion, so increasing the exposure of the tube bundle to the heating fluid.

The temperature profiles experienced by both fluids as they travel through their respective paths are shown in Figure 7.2. The heating fluid has only one pass through the vessel and its temperature declines from an initial value of T_1 to an outlet value of T_2. Meanwhile, the cool fluid enters at a temperature of t_1 and, during the first pass, proceeds down to the floating head where it assumes the temperature t_l. During the second pass its temperature rises further to the exit value of t_2. As a result, a portion of the cool fluid is flowing in parallel with the heating fluid while the remainder is flowing counter to it. This means that, in a two-pass heat exchanger, the temperature differences between the heating and cool fluid are varying

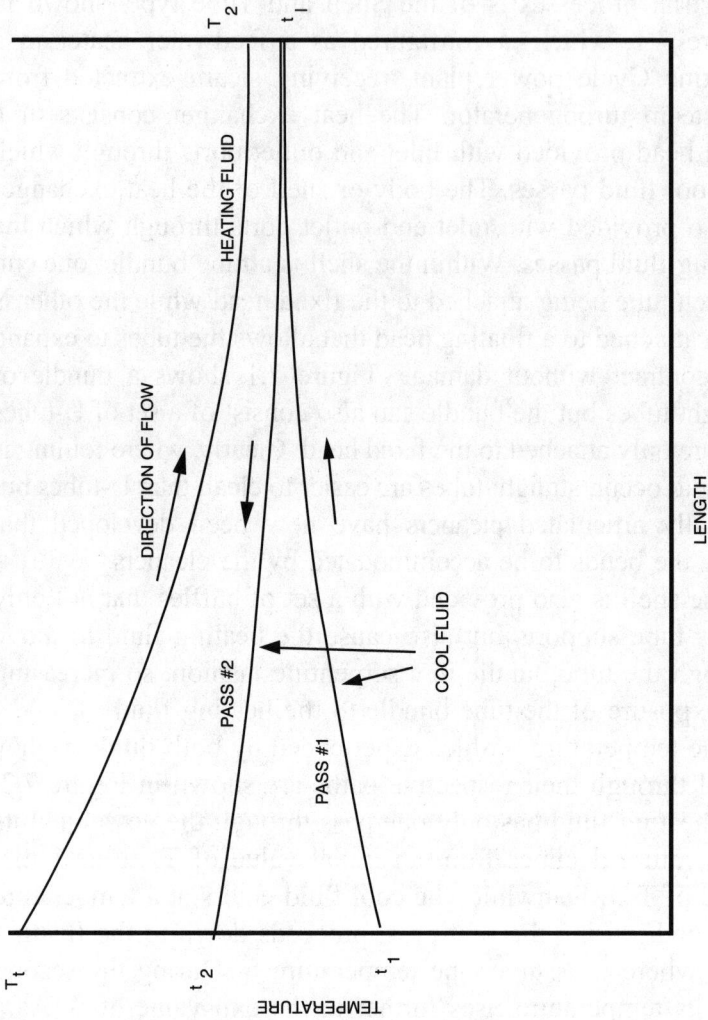

Figure 7.2   Temperature relationships for the two passes of the heat exchanger in Figure 7.1

along the length of the section and this has an effect on the performance calculations.

Monitoring the performance of a process heat exchanger is important to the economics of the process in that performance of the whole process can be reduced if the tubes become fouled. In the case of a waste fluid stream, fouling will result in reducing the amount of heat that can be recovered, so increasing the amount of energy that has to be supplied to the process. In cases of severe fouling, the reduced capacity of the heat exchanger can even affect the process production rate, again impacting the economics of plant operation.

The design procedure for shell-and-tube heat exchangers is outlined in detail in both the HEI (1980) and TEMA (1988) Standards. The purpose of the following section is to present the basic thermodynamic principles involved in heat exchanger performance monitoring, as contained in these standards.

7.1.1 Fundamental Thermodynamic Relationships—Shell-and-Tube Heat Exchangers

This discussion of the fundamental thermodynamic relationships that determine the behavior of a heat exchanger will examine those displayed by the shell and tube heat exchanger depicted in Figure 7.1. Let:

W_1 = Mass flow of hot fluid flowing on the outsides of the tubes lb/h

W_2 = Mass flow of cold fluid flowing on the insides of the tubes lb/h

C_{p1} = Specific heat of hot fluid flowing on the outsides of the tubes

C_{p2} = Specific heat of cold fluid flowing on the insides of the tubes

T_1 = Temperature of hot fluid entering °F

T_2 = Temperature of hot fluid leaving °F
t_1 = Temperature of cold fluid entering tubes °F
t_2 = Temperature of cold fluid leaving tubes °F
A = Surface area ft^2
U_{eff} = Effective heat transfer coefficient BTU/(ft^2.h.°F)
Q = Heat transferred from hot to cold fluid BTU/h
LMTD = Log Mean Temperature Difference °F
F = LMTD Correction Factor

Heat lost by hot fluid equals heat gained by cold fluid or:

$$Q = W_1 C_{p1}(T_1 - T_2) = W_2 C_{p2}(t_2 - t_1) \qquad (7.1)$$

And heat transferred can be calculated from:

$$Q = U_{eff}.A.LMTD \qquad (7.2)$$

From which the effective heat transfer coefficient for a single-pass heat exchanger can be calculated, thus:

$$U_{eff} = Q/(A.LMTD) \qquad (7.3)$$

7.1.1.1   *Log Mean Temperature Difference (LMTD)*

The log mean temperature difference for a single-pass heat exchanger is usually defined as follows:

Let ΔT_{high} = Higher terminal temperature difference °F
ΔT_{low} = Lower terminal temperature difference °F

In the case of the shell-and-tube heat exchanger:

$$\Delta T_{high} = T_1 - t_2 \qquad (7.4)$$

and

$$\Delta T_{low} = T_2 - t_1 \qquad (7.5)$$

while

$$LMTD = \frac{\Delta T_{high} - \Delta T_{low}}{\log \frac{\Delta T_{high}}{\Delta T_{low}}} \qquad (7.6)$$

7.1.1.2 LMTD Correction Factor F for a Two-Pass Heat Exchanger

In a two-pass heat exchanger, due to the presence of the baffles and the serpentine flow path, and the fact that a portion of the cold fluid is flowing in parallel with the heating fluid while the remainder is flowing counter to the heating fluid, a correction factor F has to be applied to the basic value of LMTD calculated from Equation (7.6).

Underwood (1934) first derived the expression for the correction factor F, which must be used to adjust the LMTD calculated in Equation (7.6), so as to reflect the parallel/counterflow condition; and a later version was developed by Nagle and coworkers (Nagle 1933, Bowman 1940). A derivation is included in Kern (1990), but the form given in the next equation and included in Hewitt et al. (1994) is the one presently used by HEI and TEMA for a double-pass heat exchanger and may be stated as follows. R is the ratio of the hot side temperature drop divided by the cold side temperature gain, thus:

$$R = \frac{T_1 - T_2}{t_2 - t_1}$$

and

$$P = \frac{t_2 - t_1}{T_1 - t_1}$$

then:

$$F = \frac{\sqrt{R^2 + 1} \times \ln[(1 - P)/(1 - PR)]}{(R - 1)\ln\left\{\dfrac{2 - P[(R+1) - \sqrt{R^2+1}]}{2 - P[(R+1) + \sqrt{R^2+1}]}\right\}} \qquad (7.7)$$

P is also known as the *temperature effectiveness*. The values of F for various values of R and P, as calculated using Equation (7.7), are plotted in Figure 7.3, which closely resembles the equivalent figures in the HEI and TEMA standards for a two-pass heat exchanger. Other derivations developed by Nagle and coworkers [Nagle 1933, Bowman 1940] show equations similar to Equation (7.7) for the values of F for heat exchangers with other configurations.

Thus, for a multi-pass heat exchanger, Equation (7.2) becomes:

$$Q = U_{eff} . A . F. \text{LMTD} \qquad (7.8)$$

From which the effective heat transfer coefficient for a multi-pass heat exchanger can be calculated, thus:

$$U_{eff} = Q/(A . F. \text{LMTD}) \qquad (7.9)$$

7.1.1.3   Design Heat Transfer Coefficient–U_{des}

In a shell-and-tube heat exchanger, both sides of the tubes that provide the heat exchange surface may be prone to fouling. In order that the design amount of heat should still be transmitted between the two fluids even when fouling occurs, fouling resistances are added to the design thermal resistance of the heat exchange surface, and it is this total resistance that is used to calculate the design surface area. Recommended values of these fouling resistance allowances are contained in the TEMA standard (1988) and the allowances are also stated in the design data sheet provided by the heat exchanger vendor.

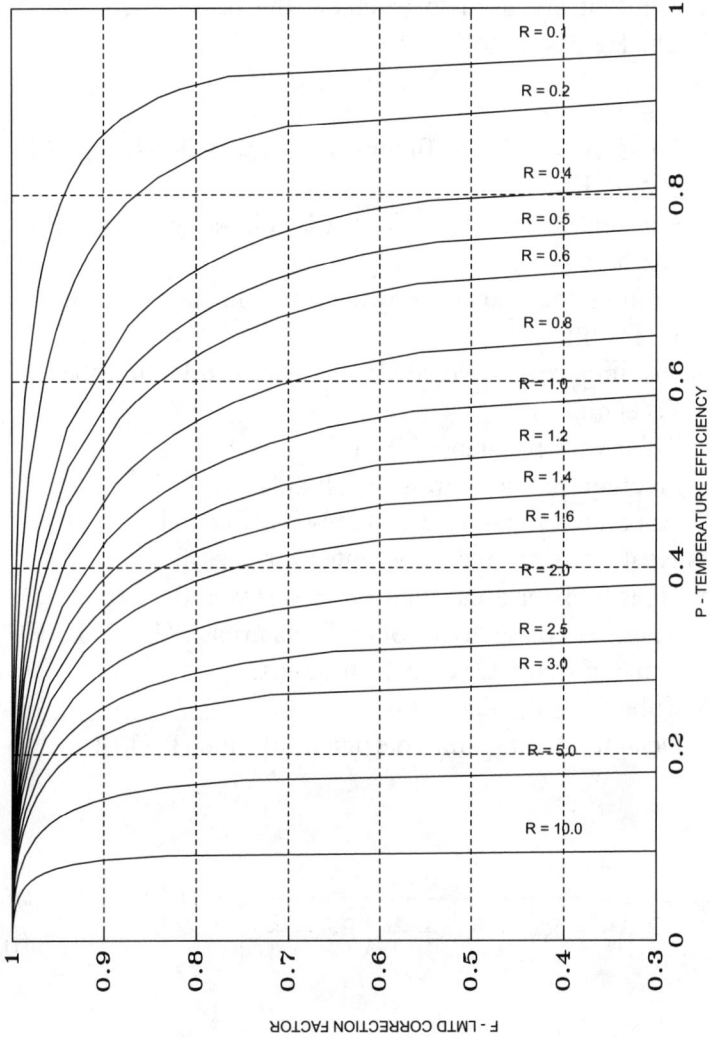

Figure 7.3   LMTD correction factors for two-pass heat exchanger

Of course, these fouling allowances cause a clean heat exchanger to perform slightly better than expected when operating under the specified design conditions. The set of thermal resistances that are used to calculate the design heat transfer coefficient H_{des} is as follows:

Let

h_{fo} = Nusselt film coefficient of shell-side fluid BTU/ (ft^2.h.°F)

h_{fi} = Nusselt film coefficient of tube-side fluid BTU/ (ft^2.h.°F)

R_{fo} = Fouling resistance allowance on outside of tubes °F/ (BTU.h.ft^2)

R_{fi} = Fouling resistance allowance on inside of tubes °F/ (BTU.h.ft^2)

R_w = Tube wall resistance °F/(BTU.h.ft^2)

T_m = Average temperature of shell-side fluid °F

t_m = Average temperature of tube-side fluid °F

A_o = Heat transfer area, tube outer surfaces ft^2

A_i = Heat transfer area, tube inner surfaces ft^2

k = Thermal conductivity of wall material BTU/(ft^2.h.°F)

d = Outside diameter of bare tube ins.

th = Tube wall thickness ins.

U_{des} = Design heat transfer coefficient of tube BTU/(ft^2.h.°F)

Then

$$U_{des} = \frac{1}{\left[\frac{1}{h_o} + R_{fo} + R_w + \frac{A_o}{A_i} \left(R_{fi} + \frac{1}{h_i} \right) \right]}$$

(7.10)

7.1.1.4   Tube Wall Resistance—R_w

It is first necessary to calculate the tube mean metal temperature that, for a bare tube, is calculated from Equation (7.11):

$$t_{mean} = T_m - \left[\frac{\left(\frac{1}{h_o} + R_{fo}\right) + \frac{R_w}{2}}{\left(\frac{1}{h_o} + R_{fo}\right) + R_w + \left(R_{fi} + \frac{1}{h_i}\right)\left(\frac{A_o}{A_i}\right)} \right]$$

$$(T_m - t_m) \tag{7.11}$$

The thermal conductivity corresponding to the value of t_{mean} should be the value used in the Kern Equation (7.12) (Kern (1958)) to calculate the wall thermal resistance of bare tubes:

$$R_w = \frac{d}{24k}\left[\ln\left(\frac{d}{d - 2th}\right)\right] \tag{7.12}$$

7.1.1.5   Fouling Resistance Allowances–R_{fi} and R_{fo}

The TEMA Standards (TEMA (1988)) contain tables showing the recommended fouling allowances for a large number of common industrial fluids that should be included in the heat transfer calculations and these tables should be referred to.

7.1.1.6   Tube Side Film Heat Transfer Coefficient–h_i

The Dittus-Boelter equation is usually used to estimate the tube side film heat transfer coefficient, as a function of the corresponding Reynolds and Prandtl number, as follows:

$$h_i = 0.023 Re^{0.8} Pr^{0.4}$$

or in its expanded and modified form:

$$h_i = 0.023 \left(\frac{k}{D_i}\right)\left(\frac{\rho D_i V}{\eta_b}\right)^{0.8}\left(\frac{C_p \eta_b}{k}\right)^{0.4}\left(\frac{\eta_b}{\eta_f}\right)^{0.14} \tag{7.13}$$

where

D_i = Inside diameter of tube ft
V = Fluid velocity ft/h
ρ = Fluid density lb/ft^3
C_p = Fluid specific heat BTU/(lb.$^\circ$F)
η_b = Fluid viscosity at bulk temperature lb/(h.ft)
η_f = Fluid viscosity at film temperature lb/(h.ft)

Kern (1958) and TEMA (1988) both include tables of properties for a number of common industrial fluids. Kern (1958) also includes several examples of procedures for calculating the value of h_i for a number of heat exchanger configurations.

7.1.1.7   Shell Side Film Heat Transfer Coefficient–h_o

The fluid flow path on the shell side is made complicated by the presence of baffles that cause the fluid to flow across the tubes. There is also a certain amount of leakage between the tubes and the baffles and there is some bypassing of fluid around the gap between the tube bundle and the shell. Kern (1958) developed a form of Equation (7.13) for estimating the shell side heat transfer coefficient, based on the extensive analysis and correlation of test data:

$$h_o = 0.36 \left(\frac{k}{D_e}\right) \left(\frac{\rho D_e G_s}{\eta_b}\right)^{0.55}$$

$$\times \left(\frac{C_p \eta_b}{k}\right)^{0.33} \left(\frac{\eta_b}{\eta_f}\right)^{0.14}$$

(7.14)

where

D_e = Equivalent diameter ft.
G_s = Mass velocity lb/(h.ft^2)

However, the estimation of the shell side heat transfer coefficient h_o is a very complicated process, especially where condensation is involved. Fortunately, computer programs are available for these calculations and should be used.

Alternatively, the information contained in the heat exchanger design data sheet provided by the vendor may be referred to. This information was, presumably, verified during the equipment acceptance test and states the overall heat transfer coefficient of the clean heat exchanger as well as the fouling resistances included for both the shell and tube sides. An example of a standard data sheet is contained in HEI (1980).

7.2 Heat Exchangers in the Alumina Extraction Process

Heat exchangers are commonly included in the flow sheets for industrial processes to recover heat from recycled fluids. However, the location in the flow sheet of such heat exchangers often makes their condition critical to the operation of the process as a whole. Fouling can affect the flow capacity of the process streams in which they are located, so affecting production rate. Fouling can also affect the heat transfer capacity of the exchanger, reducing the amount of heat recovered. Often the process requires this heat to be replaced. If steam generated in fossil-fired boilers is the source of this replacement heat, clearly the economics of process operation will be detrimentally affected.

7.2.1 The Fundamental Chemistry of the Bayer Process

The Bayer Process is an example of an industrial process in which the heat exchangers play a vital role. The Bayer Process was designed to recover alumina from bauxites and the flowsheet can take several forms. One version of the process is shown in Figure 7.4 and includes recovery of alumina from the Red Mud that might otherwise be rejected: while the simpler version shown in Figure 7.5 does not. The process begins by grinding the bauxite to a minus 10 mesh and it is then mixed with lime and soda ash, together with a stream of the mother liquor returned from the process. The chemical reactions are as follows:

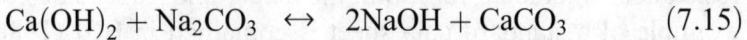

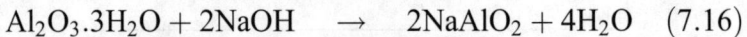

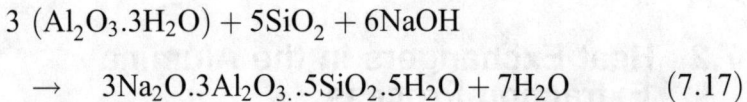

$$Ca(OH)_2 + Na_2CO_3 \quad \leftrightarrow \quad 2NaOH + CaCO_3 \qquad (7.15)$$

$$Al_2O_3.3H_2O + 2NaOH \quad \rightarrow \quad 2NaAlO_2 + 4H_2O \quad (7.16)$$

$$3\,(Al_2O_3.3H_2O) + 5SiO_2 + 6NaOH$$

$$\rightarrow \quad 3Na_2O.3Al_2O_3..5SiO_2.5H_2O + 7H_2O \qquad (7.17)$$

In Equation (7.15), the lime and soda ash first react together to form soluble caustic soda and insoluble calcium carbonate. In Equation (7.16), the alumina tryhydrate contained in the bauxite ore reacts with the caustic soda to form soluble sodium aluminate. Equation (7.17) reflects a further reaction between the alumina tryhydrate, caustic soda and the silica in the bauxite. This reaction forms insoluble sodium aluminum silicate that reports to the "red mud" stream. Note that the mother liquor returned from the process also contains caustic soda and this has to be taken into account when calculating how much soda ash and lime must be supplied to satisfy all three equations.

7.2.2 Bayer Process with Red Mud Processing

The flowsheet of Figure 7.4 starts in the top left hand corner. The bauxite received from the mine is first ground and stored. Lime and Soda Ash are also stored in bins and, periodically,

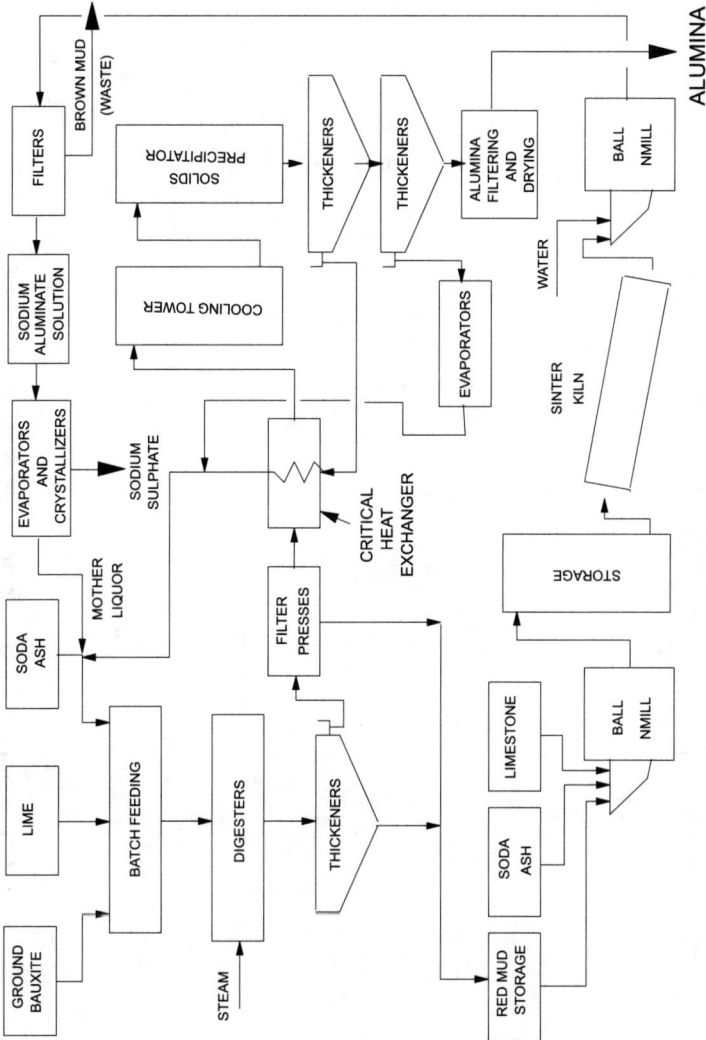

Figure 7.4 Bayer process flow sheet

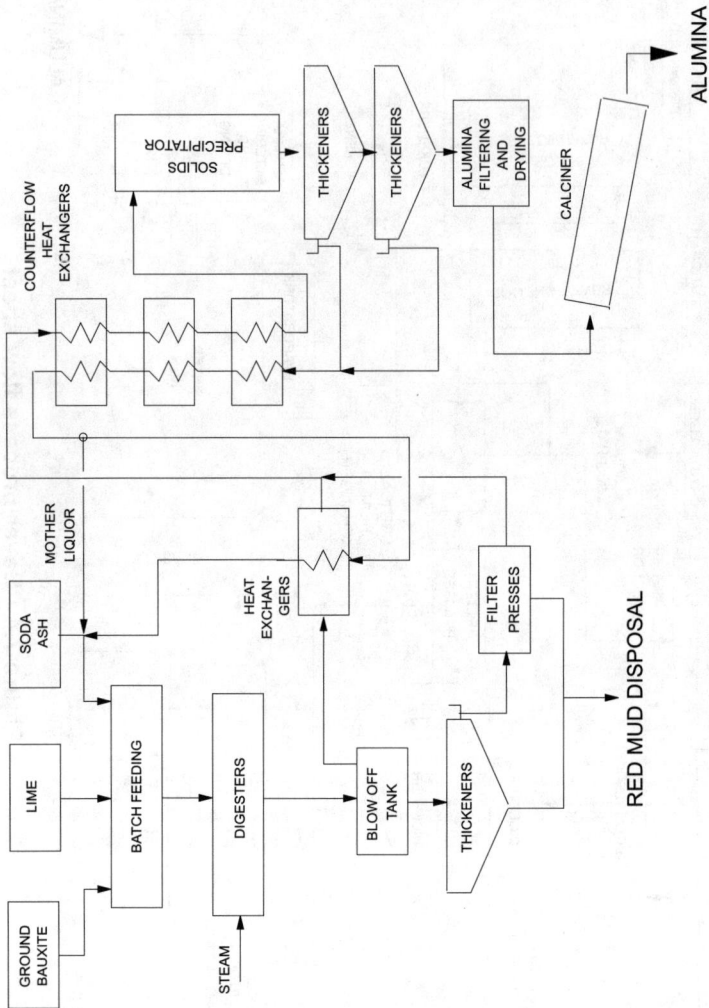

Figure 7.5   Bayer process flow sheet-ALCOA

batches of these three materials in appropriately calculated proportions are batched and then fed to one of the digesters. There are usually several batch digesters in a plant and their operation is staggered so that the rest of the system experiences a sensibly continuous process.

The reaction within the digester is aided by a supply of process steam. The boilers generating plant steam may supply it first to steam turbogenerators so as to reap the benefits of cogeneration. The process steam is obtained by extraction for the turbogenerator, which may or may not be equipped with a condenser. For instance, if the turbine is of the backpressure type, no condenser will be present.

When the process time for a batch has expired, the batch is discharged into a blow tank and the mixture then passed to thickeners. The overflow from this thickener contains the soluble sodium aluminate, the liquor being passed through filter presses and its heat then recovered in heat exchangers. The liquor is then cooled and received in a solids precipitator. The discharge from the precipitator is received by additional thickeners, the heat in the overflow from the first being recovered in the heat exchangers already mentioned; while the overflow from the second is first evaporated before being returned to the batch feeding system. The *underflow* from these thickeners is filtered and dried, the dried product being powdered alumina that is then stored for export.

The underflow from the thickeners associated with the batch digesters, as well as the residue filtered out from the liquor, is known as "red mud", a compound that consists largely of sodium aluminum silicate so it is still rich in aluminum. The "red mud" is stored so that this source of aluminum can be recovered in a separate process, shown on the lower left-hand side of the flow sheet of Figure 7.4.

After the red mud is mixed with appropriate proportions of soda ash and limestone, the mixture is passed to balls mills where it is ground to a minus 10 mesh and then stored. This ground product is then passed to a sinter kiln in which the associated chemical reactions occur, the sintered product being fed with water to a ball mill to convert the sinter to a slurry. The slurry is then filtered, the residue consisting of "brown mud", the waste rejected from the process; while the filtrate contains high concentrations of sodium aluminate in solution together with sodium sulphate. The latter can be recovered separately after evaporation and crystallization while the mother liquor containing the sodium aluminate is passed back to the batch feeding system.

In this flow sheet, the critical heat exchanger is provided to preheat the overflow from the primary thickeners as it passes to the alumina precipitation process, by recovering heat contained in the liquor being recycled from the precipitation process back to the batch feeders. Clearly, depending on pump capacity, excessive tube fouling could limit the flow of recycled liquor and so affect production rate. However, taking the heat exchanger out of service in these circumstances will require the heat not now recovered to be replaced with live steam.

7.2.3   Bayer Process with Red Mud Rejection

This flowsheet of Figure 7.5 is somewhat simpler than that shown in Figure 7.4 in that it does not include any subsystem associated with "red mud" processing. In this form of the process, the overflow from the thickener is passed through a set of heat exchangers before entering the solids precipitator. The discharge from the latter is received in two thickeners, the underflow from which contains a high concentration of alumina. This underflow is filtered, dried and calcined to produce alumina in the form of a fine white powder, which is then shipped to storage.

This flow sheet shows three counterflow heat exchangers arranged in series and provided to recover some of the heat from the liquor recycled from the alumina precipitation process stage. Another heat exchanger is provided to recover heat from the fluid stream from the blowoff tank to the precipitation process. Again, tube fouling could cause a reduction in liquor recycle rate.

7.3   Modern Heat Exchanger Maintenance Procedures

Putman's first book titled, *Steam Surface Condensers: Basic Principles, Performance Monitoring and Maintenance*, published ASME Press, New York, NY, explains successful experiences in operating water cooled condensers and balance of plant heat exchangers in fossil fuel and nuclear power plants. The methods and results are real world experiences and have withstood the test of time in a very competitive industry.

Industrial power generation plants and industries having shell and tube water cooled steam surface condensers and balance of plant heat exchangers could realize similar successful results. In addition to the tube cleaning methods in Putman (2001) and as shown in Chapter 3 of this book, unique tube cleaning methods are also used in process industry heat exchangers.

7.3.1   Hydrodrilling Heat Exchanger Tubes

Hydrodrilling consists of water driven pneumatic powered mechanical drill that drills through deposits in a completely clogged tube in a heat exchanger tube bundle on site. Water drives the drill into the deposit. Air drives the rotating drill bit. The drill bit cuts the deposit.

The rotating drill bit drills and flushes out the deposit in wet drilling operations or, the rotating drill bit drills out the

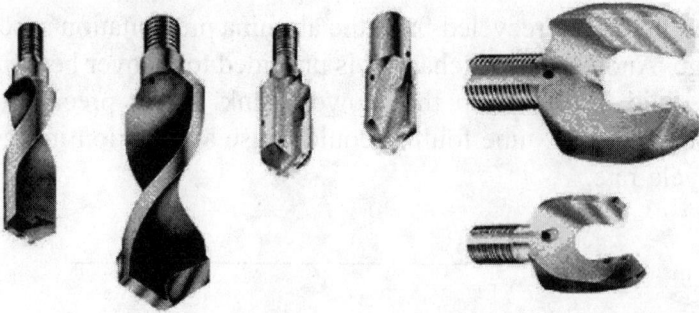

Figure 7.6   Drill bits for heat exchanger tube cleaning

deposit dry. Enclosed systems can be used to keep liquids off the floor.

Specific families of drill bit designs each having a specific family of sizes have been developed to cover numerous type plugging deposits in various industries. See Figure 7.6 and Figure 7.7 for details. Examples of deposits removed include very hard, moderate hard and soft.

Hydrodrilling can be performed in horizontal tubes or vertical tubes. Hydrodrilling can be done at the installed site of the heat exchanger tube bundle. And, hydrodrilling can be

Figure 7.7   Drill brushes for heat exchanger tube cleaning

done in a separate work area for removed tube bundles. One tube is drilled at a time. Tubes up to 40 feet long can be drilled.

7.3.2   Hydrodrilling and High Pressure Water

Hydrodrilling is significantly different then hydrolancing with high pressure water. Hydrolancing uses water pressure alone. Hydrodrilling uses a cutting drill bit and water pressure. High pressure water is usually used to clean the outside surfaces of the tube bundles.

7.3.3   Heat Exchangers with U-Tubes

U-tube heat exchangers have been effectively cleaned mechanically using specially designed tube cleaners. The mechanical tube cleaner is shot through the tube using water pressure at 300 psig. These tube cleaners are similar to those included in Chapter 3.

7.3.4   Process Heat Exchangers

Process heat exchangers having product inside the tubes are always a site specific problem. It requires a site specific solution.

- The solution could be just repeated hydrodrilling.
- The solution could be a combination of hydrodrilling to significantly open up the tube to as new condition, and monitoring the build-up to shoot the tube mechanically as needed to prevent complete tube plugging.
- The solution could be to simply use the family of mechanical tube cleaning methods shown in Chapter 3.

Chapter 8
Demand Side Management and
Electrical Load Shedding

The shedding of discrete electrical loads has been an important part of demand-side management, a policy adopted by many utilities prior to deregulation in order to delay the need for the construction of new power plants. The uncertainties surrounding the future form that deregulation will take, balanced against the present realities of current electrical tariff structure, have prompted industrial plants to take a renewed interest in reducing electrical power costs by the intelligent shedding of electrical loads. Experience has shown that managing the operation of the electrical loads within a plant by taking advantage of the terms of the current utility contract can often save between 5% and 15% of the total utility bill. This chapter examines the basic principles of electrical load shedding advanced by Putman (1975) and discusses some of the major considerations that should be incorporated in the design of the system. Of course, the increasing popularity of various forms of distributed generation will affect future generate/buy decisions but the basic principles of load shedding would still apply.

In addition to the use of intelligent load shedding to control purchased power costs, load shedding is also used to maintain the stability of the process in the event of the unexpected tripping of electrical generation or transmission equipment. Clearly, the ability to maintain product quality during such transients, or the reduction in the amount of product that has to be rejected by quality control, can have important economic consequences to the plant. The principles involved in this other form of load shedding are also discussed in this chapter.

8.1 Demand Control and Electrical Load Shedding — Fixed Window

Electrical tariffs are usually composed of two parts: (a) the cost of energy during an agreed time period, the unit energy cost falling with an increase in the consumption of the amount of energy consumed and (b) a demand charge penalty. Demand is defined as the number of kilowatt-hours of electrical energy consumed during a period divided by the length of the contractual time window. Let:

D = Electrical demand   Kw
t = Length of demand period   Minutes
E = Electrical energy consumed   Kwh

Then

$$D = E/(t/60) \tag{8.1}$$

Clearly, the reduced cost of energy with increasing consumption encourages the user to maximize the load factor experienced during each period. However, utilities are interested in maintaining the load on their systems within their present operating limits, in order to avoid having to make unplanned investments in new transmission or generation equipment. Thus the contract with the utility specifies the agreed maximum demand that may be drawn from the utility tie-line during a contractual time window and, to discourage this limit being exceeded, a demand in excess of this amount is subject to a specified penalty. Thus two important principles that must be incorporated in the design of a load shedding demand control system are:

- Ensure that the demand during any period is below the agreed demand limit
- Assuming that all energy consumed increases the value of the product, maximize the energy consumed during each period.

In many contracts different electricity rates are used for on-peak and off-peak periods, while the demand limit may also be raised during off-peak periods and a demand control system must be able to respond to the latter.

Figure 8.1 illustrates the operating envelope within which the power consumption trajectory must fall. The ordinate is calibrated in kilowatthours even though the demand limit is stated in Kw, while the abscissa represents the time into the period. The mean trajectory is shown by line XY but this does not maximize the amount of energy consumed. The real bounds of the operating envelope are formed by the base (or uncontrolled) load, which is present during every period; together with a line that represents the sum of the base load plus all controllable loads if the latter are not switched off during a period. Trajectories that generally follow tangents to these two bounds (line XAY) are ideal. Other trajectories such as XBY and XCY are acceptable with regard to not exceeding the demand limit but, because equipment has been switched off when it could have been usefully employed, have not maximized the amount of energy that could have been consumed.

Originally there were two signals sent out by the local utility associated with power demand control:

- A pulse representing the beginning of each demand period, transmitted by the utility every 15 or 30 minutes

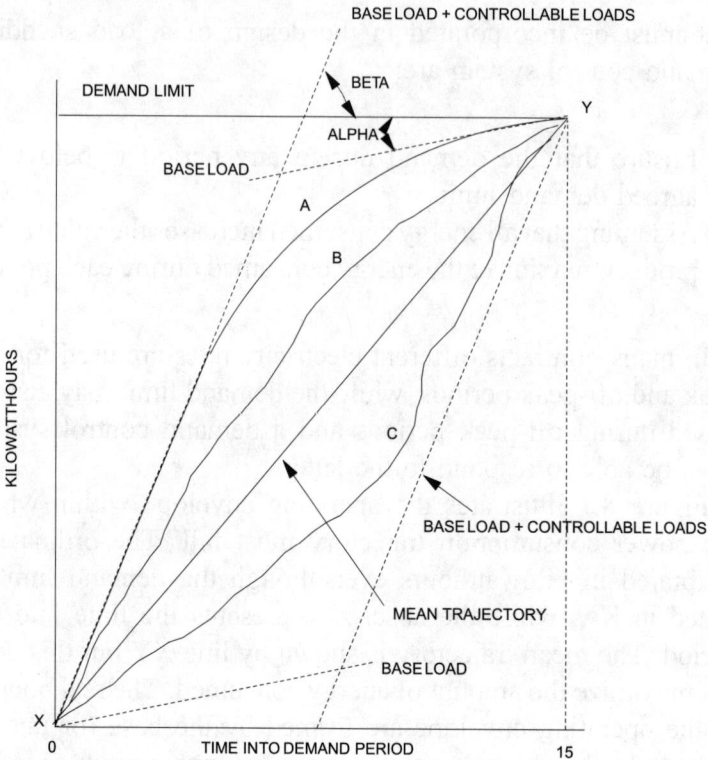

Figure 8.1   Total load trajectories

- A pulse indicating that the disk of the kilowatthour meter measuring tie-line power flow had completed a revolution. Knowing the number of kilowatthours per revolution of the disk and having digital means available for storing the time each pulse was received, it is possible to calculate the rate at which power was flowing from the tie-line. Thus:

Let KWH = number of kilowatt hours
              per revolution of the disk   Kwh

Δt = time difference between two
   consecutive pulses   secs

KW = power flow from the utility
   tie−line   Kw

Then

$$KW = KWH/(\Delta t/3600) \tag{8.2}$$

More recently, kilowatthour meters are being provided with magnetic tape cassettes for recording the kilowatthour meter pulses, these tapes being changed once per month. They are then analyzed off-line to establish the maximum demand calculated during any 15- or 30-minute period during that month. This is known as a sliding window and increases the complexity of the control strategy. It will be discussed further in Section 8.2 below.

8.1.1.1 Predicting Demand Error at the End of a Period

Consider Figure 8.2, in which the trajectory at time t (point E) shows that D kilowatthours have been consumed, while the rate of tie-line power consumption KW can be calculated from the kilowatthour meter pulses, using Equation (8.2). If conditions were to remain unchanged until the end of the period (point F), then predicted demand error FY in kilowatthours can be calculated from:

$$FY = D + KW * ((15 - t)/60) - Y \tag{8.3}$$

Clearly, control decisions can be based on the magnitude of FY.

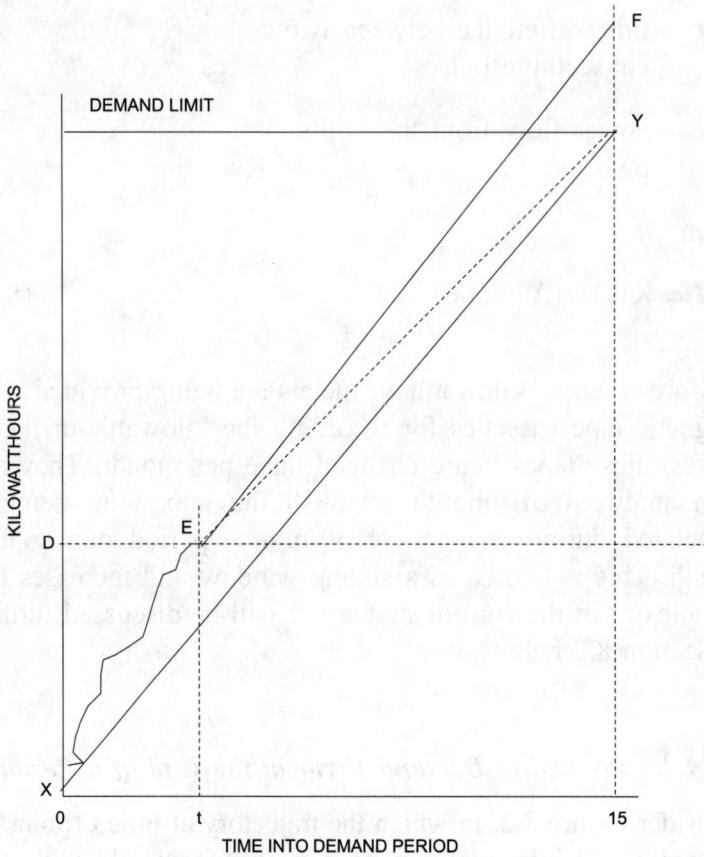

Figure 8.2   Demand limit prediction

8.1.1.2   *Types of Electrical Load*

Loads that consume energy are of several types:

- Lighting and space heating loads that are normally relatively constant and usually form part of the base load.
- Loads with short run-up times and reasonable starting currents
- Loads with extended run-up times and large starting currents

- Loads with their own on/off controller such as air or ammonia compressors, air conditioners, etc.
- Large loads of short duration and relatively infrequent occurrence, e.g. test loads
- Arc furnaces and similar industrial process equipment where production penalties are paid for extended shutdown periods.
- Loads associated with safety requirements, e.g. possibility of build up of toxic atmosphere if vent fans shut off for too long a period

For the purposes of power demand control these loads may be assigned to one of several categories, the assignment being very site-specific:

- **Base loads**, such as lighting or space heating, that are excluded from the control system but consume power during every period. Details about the individual loads do not need to be available but an estimate of the total base load should be known. Because these loads may be switched on or off by the process, the base load is not constant but contains some noise.
- **Inhibited loads**. Among these may be included intermittent loads associated with tests or having high starting currents. A logical variable is assigned to such loads to indicate that they are available to be started, but only when it is appropriate after considering the history of the demand period and ensuring that the control objectives will be met. They are usually started at the beginning of a demand period, switched off after a predetermined duration and inhibited from being switched on again until the beginning of the next period.
- **Critical loads**. Loads with their own on/off controller, such as air or ammonia compressors, air conditioners, etc. that are not normally part of the demand control system.

However, they can be switched off if the demand limit would otherwise be exceeded, but only after all available sheddable loads have already been switched off.

- **Sheddable loads.** Sheddable loads are defined as loads that will be switched on if they have been off longer than an assigned off time; and will remain on until at least a stated on/ off time ratio has been exceeded. They include loads such as ventilation fans that may be switched off for limited amounts of time without seriously impacting process performance. The data set required for each sheddable load is as follows:

  - Change in power consumed when load switched off
  - Maximum "off" time allowed – e.g. ventilation fans should not be switched off too long if there is a danger of toxic fumes accumulating
  - Minimum on/off ratio – used to check whether load can be switched on again, subject to maximum number of starts not having been exceeded
  - Maximum number of starts permitted per hour
  - Priority within load group
  - Subpriority within the load group
  - Logical variable indicating that the load is currently permitted to participate in the control strategy

However, demand limit control does not always have to be accomplished only by switching loads on and off. The same result can be obtained by other alternatives. For instance, fan loads can be reduced to some 20% of normal by closing the inlet vanes or dampers by means of a servomotor. This operation can be performed relatively frequently and for short periods; for example, towards the end of a demand period as a fine trim, so avoiding the limited number of starts per hour allowed for large motors which would otherwise cause the windings to be damaged. When the load is associated with eddy current couplings or pneumatic clutches, the mechanical loads can be disconnected from their motors. With air compressors, the demand control system can override the control of the inlet valves and hold them open when appropriate. Where these alternatives are available, they reduce the cost of plant equipment maintenance.

8.1.1.3 Principal Features of a Demand Control System Strategy

The previous section referred to two of the fundamental principles of a demand control strategy, namely:

- Ensure that the demand during any period is below the agreed demand limit
- Assuming that all energy consumed increases the value of the product, maximize the energy consumed during each period.

The penalty incurred when the demand limit is exceeded can be quite severe and may continue for several subsequent months. A control strategy can be adopted that will reduce the likelihood of this happening, some of the features being as follows:

- Divide the period into three zones and choose a different demand limit target for each zone. For instance, during the first few minutes of each period, there is little information available for long-term predictions. During this **first period**, inhibited loads might be switched on, thus allowing more time to correct for any excessively high demand errors that might be predicted later in the period. A variable dead-band approach might also be considered in which only large errors will be responded to during the first part of each period (See Figure 8.3). This tends to reduce the number of starts, too-frequent starts tending to damage motor windings and increase the cost of maintenance.

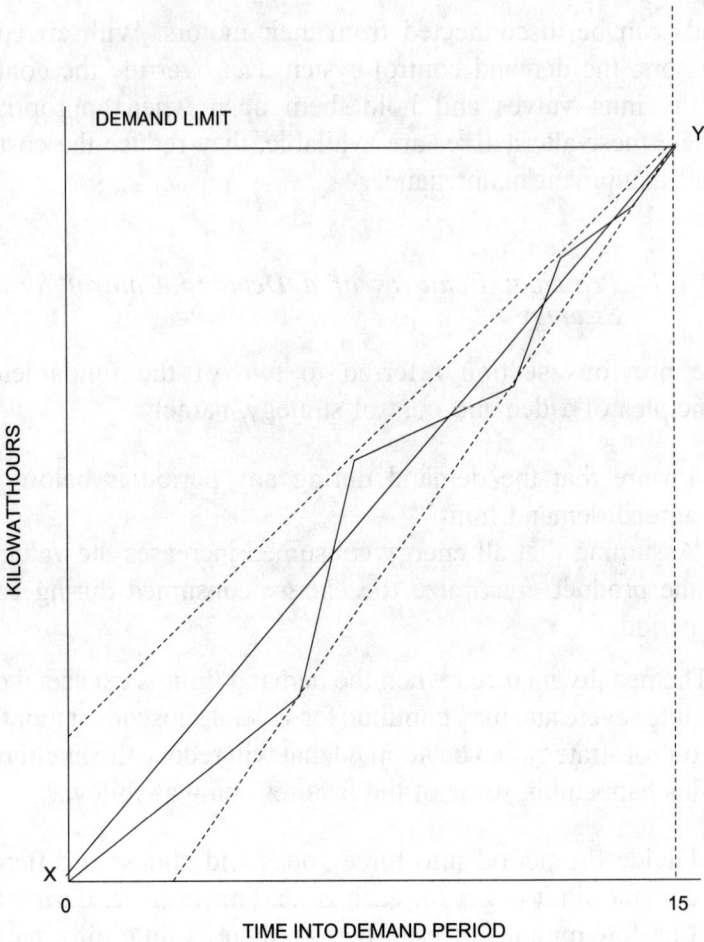

Figure 8.3   Variable control dead band

- During the **second period**, the target demand limit can be reduced below the contractual value. While this will cause a reduction in total energy consumption below the maximum possible, it also provides some demand error cushion when entering the third period.
- During the **third period**, raise the demand target to its contractual value.

- Sheddable loads will be switched on or off in a sequence that is determined by the group and subgroup priority assigned to each load. Note that the control algorithm should be designed so that loads can be switched both on and off. While switching a load off safeguards the demand limit, switching a load on tends to maximize the amount of energy consumed. Meanwhile, the lower the group priority assigned to a load increases the likelihood for that load to be chosen for shedding. Thus assigning a low priority to a load means that it is less important to the process and can be switched off preferentially. Higher priority assignments mean that those loads are more important to the process and will be shed only after lower priority loads have already been switched off.
- To avoid the same load in a group always being selected as the first to be switched off, rotate the group subpriority assignments periodically.

The control algorithm should be exercised only when new information has been received (e.g. a pulse from the kilowatthour meter). The demand error FY calculated in Equation (8.3) may be used to determine the change in power ΔP needed to restore the error to zero, thus.

$$\Delta P = -1.0 * FY/((15 - t)/60) \qquad (8.4)$$

If ΔP is negative, this indicates that loads should be shed and, if positive, that loads should be switched back on.

If the predicted demand is greater than the desired value, the system examines the lowest priority load to be dropped to see whether it is on and, if so, whether there are any constraints that prevent it from being switched off. If it cannot be dropped, the next lowest load is examined until a load is found which is available for control. However, if the sheddable load found is larger than the demand limit deviation, another load is sought for switching.

As the control algorithm searches the list of sheddable loads in the sequence determined by the assigned priorities it will not only select and switch the loads to be controlled but also deduct their contribution from ΔP, continuing the selection and switching process until ΔP is reduced to zero. Negative demand deviations are handled in a similar fashion.

8.1.1.4   Demand Control Simulator

When installing a demand control system the amount by which it will reduce the demand limit is of significant interest. Due to the inherent random variations in power consumption that all plants exhibit, the control system should not be overly constrained. If the demand limit is set too low, this might occur resulting in the limit occasionally being exceeded. This must be avoided. It is clearly possible to write an off-line simulation program that uses the data base associated with the assigned set of loads and reproduces the switching logic and rules that have been incorporated in the on-line control system. Such a program can be exercised to test how low the demand limit can be set while still maintaining the control objectives. This also provides an opportunity to evaluate different combinations of load assignments as well as the effect of their constraints, even before a system is actually installed. An example of a simulator was included in Putman (1975).

8.2   Demand Control and Electrical Load Shedding—Sliding Window

When power demand control systems were first introduced in the early 1970s utilities were willing to provide not only pulses from the utility's kilowatt meter but also a system pulse indicating the beginning of a new demand period. However, a number of users found that knowing when the latter occurred allowed them to start large loads at the beginning of a period, and manage these and other loads during the remainder of the period, without causing the demand limit for that period to be exceeded. Clearly, the initial surge in current was much higher than the contractual demand limit. This technique was employed by several steel mills in the northeast to determine when to start a rolling mill and that synchronization caused significant and unexpected load changes to be imposed on local power plants. The utilities became alarmed since increases in local generation capacity had not been planned. This gaming of the terms of the demand control contract was frustrating the whole concept of demand-side management.

The response of the utilities was to suspend the transmission of the period pulse and to install new kilowatthour meters on their tie-lines equipped with cassette recorders that would record the time of each kilowatthour meter pulse. At the end of each month these cassettes were replaced and, using the sliding window concept, were analyzed off-line with the help of a computer to determine the value of the maximum demand experienced during *any* 15- or 30-minute period during the month. The absence of the period pulse eliminated the gaming of the contract and resulted in a significant change in control strategy on the part of users.

The maintenance of power demand limit during any 15- or 30-minute period now became the primary objective and the switching principles developed for the fixed window approach still applied. However, the secondary objective of maximizing energy consumption within the demand limit could no longer be implemented.

The only reasonable control strategy when demand is calculated as a function of a sliding window is to continuously maintain the power drawn from the tie-line at, or slightly below, the demand limit. The load selection and switching logic that was developed for the sheddable loads in the fixed window case can still be utilized. However, the switching on of large loads will probably have to be confined to the off-peak periods; although a simulator could be used to determine by how much the demand limit should be suppressed prior to switching on one of these loads. Clearly, the adoption of the sliding window principle by the utilities reduces the magnitude of the savings that might be realized through the practice of tie-line demand control.

8.3   Load Shedding in Response to Plant Contingencies

Many industrial plants are not only connected to the local utility tie-line but are also equipped with in-house turbogenerators. These sources of electrical energy operate in parallel, together supplying power to numerous motors, heaters and other equipment through the internal power distribution system. Figure 8.4 is a one-line diagram typical of a number of such plants.

The power from the tie-line is often brought into the plant via a step-down transformer and each load of known magnitude is discretely connected to a particular distribution bus. Circuit breakers S1 and S2 are provided at the tie-line and after the transformer respectively; while a third circuit breaker S3 is provided between the generator and the internal bus.

When all the equipment is operating properly, the total amount of power required by the process equipment is distributed between the tie-line and generator according to related economic criteria, while the bus voltages and tie-line power factor are being held within their assigned tolerances. However, such a plant can be subjected to disturbances or contingencies of various kinds; in this case, the tripping of the

TIE-LINE

S1

TURBOGENERATOR

TIE-LINE
TRANSFORMER

S3   S2

LOAD # 1   LOAD # 8

LOAD # 2   LOAD # 9

LOAD # 3   LOAD # 10

LOAD # 4   LOAD # 11

LOAD # 5   LOAD # 12

LOAD # 6   LOAD # 13

LOAD # 7   LOAD # 14

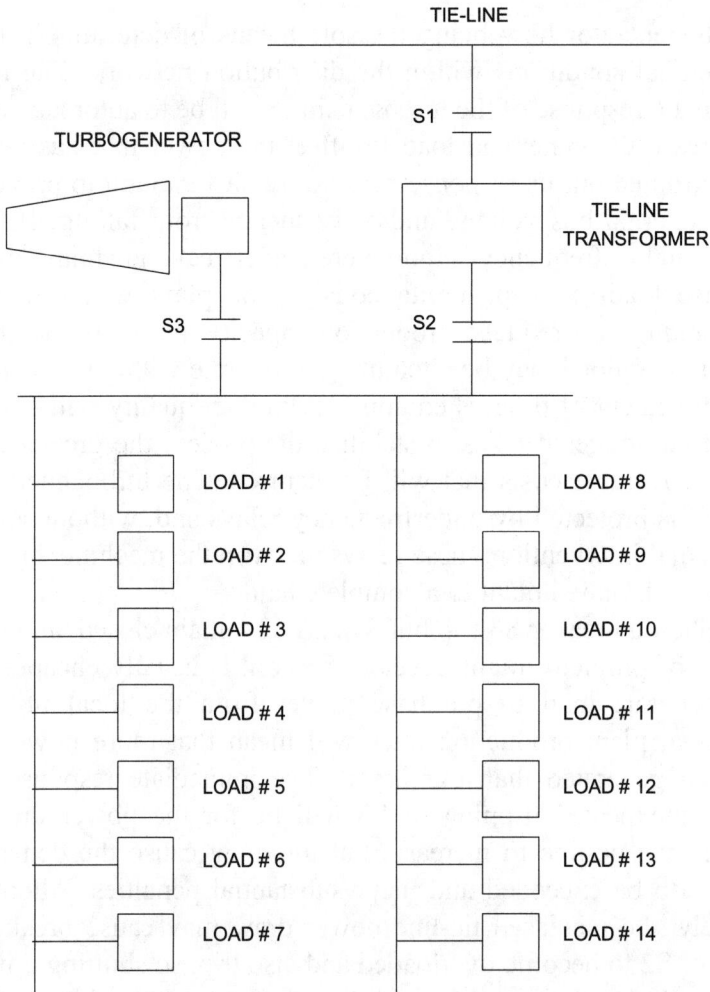

Figure 8.4   Power distribution system small industrial plant

tie-line breaker S1, some problem with the breaker S2 located after tie-line transformer, or the tripping of the breaker S3 associated with the turbogenerator.

The immediate response to the tripping of S1 or S2 is rather similar, causing the plant to become an island, the turbogenerator now being the sole means of determining the electrical conditions within the distribution network. The immediate response of the turbogenerator will be to automatically increase the generated load to offset the power that has been interrupted but its response may not be fast enough to prevent the internal bus voltage and/or frequency from falling. If the drop in bus frequency is too severe, underfrequency relays may cause loads to trip spontaneously. The plant will now be operating at a reduced production capacity and some process unit operations may be attempting to operate without their full complement of process equipment. Product quality will suffer and the longer it takes to stabilize the process the greater the unrecoverable costs that will be incurred. The turbogenerator itself is protected by underfrequency relays and, without some external intervention, these relays may trip the machine, bringing the islanded plant to a complete halt.

Should breaker S3 trip but S1 and S2 remain closed, another type of problem might develop. Since it is usually cheaper to cogenerate than to purchase power from the local utility, normal plant production rates will mean that more power is being generated than purchased. The immediate response to the unexpected tripping of S3 will be for the power drawn from the tie-line to increase: but this may cause the demand limit to be exceeded and incur substantial penalties. Alternatively, the increased tie-line power draw may cause breakers S1 or S2 to become overloaded and also trip, so shutting down the whole plant and, again, causing unrecoverable costs. Restarting the plant from a complete shutdown presents its own difficulties and some equipment may have to be cleaned and the product that is removed wasted.

An effective strategy for maintaining a greater level of process stability after a power system contingency has occurred, is to identify a set of possible contingencies and to assign the loads that will be intentionally tripped in response to each and the order in which they are to be tripped, this assignment being very site-specific. The amount of load to be shed by a particular contingency will vary so that not all the loads assigned to that contingency need to be shed on each occurrence. Furthermore, a given load may be included in the planned response to more than one identified contingency. Clearly, the set of loads assigned to a contingency will determine the intended state of the process after the contingency has occurred, as well as the state of the electrical distribution system.

Once the hierarchy of loads has been created, the normal function of the contingency response system will be to select the current set of loads to be shed in response to each contingency, should it occur. Having determined the amount of load to be shed in a given contingency, the priority list for that contingency is then examined and the set of loads is selected that will achieve the desired load reduction. For each load selected, a 'trip' logical variable associated with that load and contingency will be set, so prearming the load. When the contingency subsequently occurs, a 'trip' command will immediately be sent out to all those loads that have been prearmed for that contingency and the plant brought to the level of stability that had been planned.

Such a forced response to the tripping of either breaker S1 or S2 will be that a small number of loads will be shed: but they can be restarted almost at once at a rate that will match the response of the turbogenerator as it attempts to maintain the internal bus voltage.

The number of loads to be shed in the case of breaker S3 tripping will usually be larger. A decision can now be made as to how the plant should be operated in the absence of the turbogenerator. Perhaps only a portion of the plant can continue production, especially if the previous demand limit is to be maintained, and the associated set of loads may now be restarted. On the other hand, the economics of plant operation may permit an increase in the demand limit and a willingness to pay the associated penalties. There will usually be a different set of production equipment or loads associated with this decision and, once made, that set of loads will be switched back on.

In some plants there may be more than one turbogenerator, in which case the tripping of one will cause a smaller increase in tie-line power. However, the logic now has to ensure that at least one is operating in a synchronous mode, the selection rules having been predetermined.

In both cases, the response of the process to the unexpected tripping of a breaker has been planned together with the process recovery procedure. There will be a significant reduction in the amount of product that had to be rejected as a result of the contingency and a more controlled restoration to full production once the original problem has been rectified. Meanwhile, every effort is made to maintain the existing tie-line demand limit, if that is possible.

Note that the logic involved in these programs not only checks the magnitude of each load and the bus to which it is connected but also the quality of the data associated with that load.

8.4 Load Shed Speed and the Control System

There is an important difference between load shedding for demand side management and for plant contingencies. The difference is the speed at which the shedding must occur.

The shedding of load as means of performing demand control can be a relatively slow process. The shed times can be in seconds, because it is the average power consumption during the demand period that is important. If the power purchased in the beginning of the period is large, the estimated power usage for the period could exceed the demand limit. However, load can be shed several seconds after the estimated power usage calculation and there will be time for the average power consumption to be less than the demand limit.

When load must be shed in response to an electrical disturbance, such as loss of tie line or the tripping of an interbus breaker, the load-shed operation must be particularly reliable and fast. When there is an electrical disturbance the main goal of the load-shed system should be to guarantee that the frequency resulting from a generation load unbalance does not reach an underfrequency value too critical for the plant to survive. This means that from the time the contingency occurs, until the time the load is shed, the time period must be in milliseconds rather than seconds. Practice has shown that this time period should be less than 200 milliseconds.

Detecting a contingency and issuing a shed command in less than 200 milliseconds can be a challenge for a microprocessor based distributed control system. If the entire load shed logic and process I/O required for the load-shed system are contained in one drop the speed is usually attainable. However, often times the logic that checks for contingencies is located in one area of the plant and the loads that must be shed are located several kilometers away. This means that a shed command signal must be broadcast from one controller to another via the control systems data highway. This transfer can consume as much as 100 milliseconds. Therefore, the load-shed system should be designed with the control systems physical charactcristics in mind to ensure the logic will execute fast enough to prevent an underfrequency condition in the plant.

Chapter 9
Future Trends

9.1 Real Time Electric Power Pricing

Deregulation in the electric utility industry is forcing utilities to identify ways to maintain revenues and maximize profits (Elleson (2000)). This is also having an impact on the industrial plant that must purchase power. In the regulated environment, the price for electricity is equal to the cost to produce and distribute power plus the profit as guaranteed by regulators. In the new, deregulated, marketplace profit is equal to price minus the cost to produce and distribute the electricity.

Currently, there are two main types of pricing for industrial consumers, namely, Demand and Time of Use (TOU). With demand pricing the energy cost is comprised of two components: a demand charge based on the amount of electrical capacity the industrial plant wants to be guaranteed of having in case it needs it, and a energy charge for the actual amount of power that is used. With a demand pricing contract the customer gets the energy when he wants at a price based on the amount of power required. Under the Time of Use pricing option, the price of electricity depends on the time and day and season of the year. This is similar to the way airlines adjust fares during peak travel times. TOU contracts benefits customers that have the flexibility to shift usage to off-periods when the price of electricity is less.

Publisher's note: ASME Press is grateful to Frederick C. Huff, who generously completed Chapter 9, which was unfinished at the time of the author's death.

A utility that offers demand pricing must determine an appropriate price that will cover the average cost of supplying power. The utility runs the risk that during times when the power cost is high, its customers' total demand will be greater than was estimated when the rate was established; thereby, lowering profits. Time-of-use rates represent a step toward matching the price of electricity to the cost of providing it. However, the utility still bears some risk that its cost of production and distribution will not be covered.

An example of this occurred in the Midwest during the summers of 1998 and 1999, when wholesale electricity prices experienced peaks as high as $10,000 per MWH (Elleson (2000)). Most utility customers were insulated from these price spikes, because they had demand contracts. With demand contracts their prices were fixed regardless of the electricity demand. In order to provide power to their customers some utilities were forced to buy power at a price 100 times higher than their selling price.

Due to problems such as this utilities are now starting to offer real-time pricing. Real-time pricing is an extension of the time-of-use pricing concept, whereby the rate for electric service varies hourly depending on the actual cost of production. Under real-time pricing, rates are communicated to the customer a short time ahead of delivery, typically one day. This type of pricing reduces a utilities risk. With this type of pricing the price of electricity is linked to the cost the utility must pay to provide the power. Also, it forces consumers to try to reduce their power demand when prices are high. This reduces the amount of power the utility must purchase at high market prices.

Industrial facilities that purchase power on a real-time basis have traditionally utilized modems to access and retrieve hourly rates from the utility. However, utilities have started to make the rates available on a secure web page. As a result companies such as, Automation Applications Inc, LLC and Automated Energy have developed packages for industrial consumers to automatically retrieve the hourly electric rates. By retrieving these rates automatically the plants energy management systems can be sure they are using the latest electricity cost when determining plant loading.

9.1.1 Markets for Real-time Pricing

There is much interest in Real-time Pricing (RTP) as a competitive tool but many utilities and customers are unsure if it will benefit them. In theory RTP should be a way for utilities to provide their customers with prices that allow them both to manage loads, reduce costs and maximize profits. It is also a way for utilities to increase their competitiveness and support the retention and growth of their customer base. When a utility offers RTP it usually satisfies one of the four objectives listed below.

The first objective is load management. If a customer has high load demands during peak power periods, but has the ability to reduce his demand during these periods then not only can the customer save money but it also helps limit the power demand during these periods. However, if a customer is not able to reduce his demand during the peak periods when prices are higher then there is no impact on the peak load demand. Secondly, if a utility can offer RTP as an option it can help them keep current customers that have the ability to adjust their power usage during peak periods. Related to this is the third objective. RTP can be helpful in expanding the utility customer base. Finally, it is needed to remain competitive.

According to Weisbrod et al (1996), unless a customer meets one of the criteria listed above there is no advantage for a utility to offer RTP to the customer. In fact the utility could actually lose money by offering it. An example is a newspaper printing plant that has its highest power demand at night during off peak hours. The business may reap the benefits of lower off peak rates but never have to reduce, reschedule or shift load. Since it is the newspaper it is unlikely that it would leave the region.

A customer should not be interested in RTP unless they have at least one of the following capabilities:

1. The ability to reschedule load, and the knowledge of how to do so to take advantage of the changing power prices.
2. The capability to apply an energy management system together with a control system to effectively shift load and take advantage of RTP.
3. Availability of backup generation, knowledge of when it is appropriate to shift to it and capability for it to go online and work effectively when called upon.

Examples of interested customers in RTP are oil and gas pipelines. These customers have operations that cross more than one utility service area. While electricity costs of pumping stations can account for as much as 40% of total costs, there is also typically over-capacity at pump stations and an ability for flexible scheduling of various pump unit operations at different locations. This affords an opportunity for minimization of costs by utilizing pumps in different service areas depending on the real time power prices in effect at those locations at any given time. Other industrial businesses with plants around the country can effectively accomplish the same type of optimization by reallocating production among different plants on a daily or in some cases an hourly basis.

9.2 Neural Networks

According to Immonen (2000), RTP is forcing industrial power plants to re-evaluate their operating practices. In addition, the plants have to be able to adapt to new types of operational constraints imposed by ever tightening environmental regulations. The biggest issue is emissions from the combustion process. These are items such as nitrous oxide (NOx), sulfur dioxide (SO2), carbon monoxide (CO) etc. To be able to face the new challenges and to take advantage of the new opportunities, effective tools are necessary to plan and manage the operations of generating facilities. Mathematical models and optimization software play a very important role in accomplishing this.

Real-time optimization of a combined heat and power plant has two basic functions: to coordinate the use of plant equipment in the most economical fashion, and to optimize the internal performance of each powerhouse component.

The purpose of plant-wide optimization is to find optimal settings for unit loads, fuels, and energy purchases and sales. As the operating conditions are often subject to large and rapid variations, the optimization task needs to be performed repeatedly—every couple of minutes. This requirement is often caused by rapid energy demand variations, or by electricity price fluctuations especially when RTP arrangements are in place. To do this optimization a plant-wide energy acquisition model is needed and the optimization needs to be based on real-time power and steam demands. The plant wide model is comprised of steady-state models of the power plant components and other process equipment. This type of optimization has been discussed in chapters 4 and 5.

The internal optimization of power plant components represents another type of modeling problem. The internal behavior of process equipment tends to be much more nonlinear than the overall performance of the plant.

Combustion optimization is probably the most important application of internal optimization. A boiler typically has several controllable parameters that have significant impact on emissions and fuel economy. These parameters are interactive, and their relationship is highly nonlinear and the form of the model is unknown. Therefore, in order to be effective, coordinated model-based control techniques should be used for the implementation of combustion optimization systems. The core of the combustion optimization system is the model predicting the boiler emissions (NOx, CO etc.) In order to develop this type of model neural network technology can be used.

A neural net is made up of simple, interconnected parallel calculation elements, called nodes. Each node is capable of performing one single specific mathematical operation on the weighted sum of its input signals and to transmit the information thus processed to the nodes arranged in the following layer. Based on statistical data, a learning algorithm adjusts the weights to correlate the known inputs to the known outputs. Once a satisfactory correlation has been formed, the neural net is trained, and it is now able to predict outputs from a new set of inputs. By adding nodes and layers, extremely complex nonlinear relationships can be modeled. A typical neural network structure is shown in Figure 9.1.

The main reason to use neural network technology for empirical modeling of power plant components is that it is convenient for on-line software applications. A given neural network varies only in dimension and the value of its coefficients when trained for different systems. This is in stark contrast to traditional regression packages that require different underlying mathematical functions (e.g., logarithmic vs. polynomial) to adequately model different systems.

The inputs to the neural net model are standard measurements available from the DCS, such as fuel flows, burner tilt, air and flue gas temperatures, pressures etc. The selection of inputs should be based on a good understanding of the process being modeled. This keeps the number of inputs to a minimum. Reducing the number of input variables simplifies the model design and improves its overall reliability. One of the disadvantages of neural network technology is that it has the potential to model any inaccurate inputs. Therefore, it requires process expertise to recognize the out lying data and

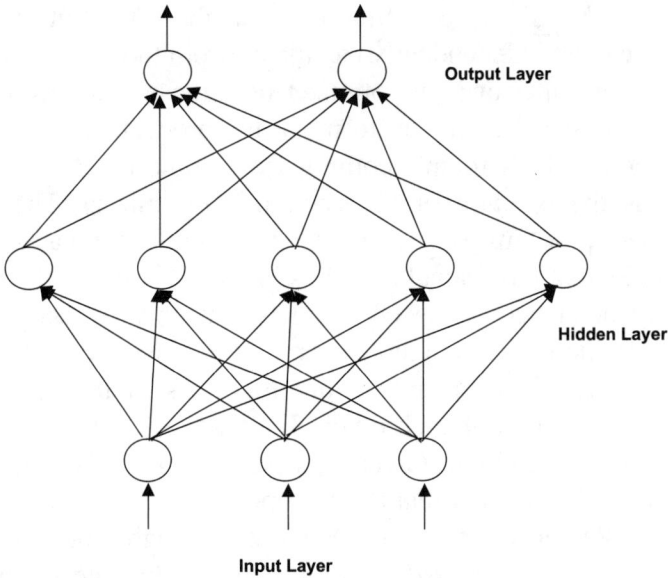

Output Layer

Hidden Layer

Input Layer

FIGURE 9.1   Feed-Forward Neural Network.

potential erroneous input data and to eliminate them from the data sets used to train the model.

A neural network based controller used for say, combustion optimization, can be thought of as being comprised of three separate sub-pieces:

Modeler
Path Optimizer (controller)
Economic Optimizer.

The modeler finds the relationships between the manipulated variables (MVs) and control variables (CVs) based on historical data. The modeler builds a model relating all MVs to all CVs. Internal to this model there are often individual models relating all MVs to each CV, one internal model for each CV. This is simply saying that the behavior of each CV is solely determined from the MVs, and is not dependent on the other

CVs. All MVs may be common to all the CVs, but the CV values can be independently calculated from those MV values.

The controller or path optimizer uses those relationships to find the best-behaved way to move the process from setpoint to setpoint (the setpoints coming from some external source such as the operator or the economic optimizer). The path optimizer first "inverts" the modeler model, forming a controller model that now calculates MVs given CV values. Note that the independence of the CV models that existed back in the modeler model no longer exists in this inverted controller model. The MVs are not simply functions of the CVs, but are also functions of each other. Simultaneous solution of the MVs is now required. Given a specification of CV setpoints from outside the controller (by the operator, say), the controller can calculate a sequence of MV values through time, for each MV that moves the system to the new steady state setpoints. There are many such MV sequences possible, but the optimization that the controller does is to find such a sequence that produces a minimum of upset (minimum number and size of the moves, etc.) for the process. The controller also makes certain that the MVs remain within the operating limits specified for them. Operating limits for the CVs may also be specified, but they are really not used in the controller since the setpoints are specified explicitly. In general, if there are more MVs than CVs in the problem, then not only can the CVs be brought to their setpoints, but some MVs can be brought to ideal resting values (again specified from outside the controller, such as by the operator).

The economic optimizer uses economic information to determine the most economical setpoints and passes those setpoints to the controller. The economic optimizer formulates an objective function consisting of the MVs and CVs multiplied by cost or revenue values. This cost objective is minimized to produce CV setpoints and MV ideal resting values that are in turn passed to the controller, which then acts on them as described above. The economic optimizer does consider and obey the operating limits for both the MVs and CVs as specified by the operator. If feasibility is not possible, operating limits for CVs begin to be violated, whereas MV constraints are hard.

If there exists a CV setpoint (or several CV setpoints) which is not to be determined economically, such as a required power output, the upper and lower operating limits for that CV should be set by the operator to be nearly equal, just above and below the desired setpoint. The economic optimizer then won't "play" with that CV setpoint, but all the economics of its related MVs and the other CVs (and their MVs) will be accounted for. The CV setpoints and MV ideal resting values thus calculated by the economic optimizer and passed to the controller will then be the most economic values given the values of the "locked" CVs.

According to Swirski and Williams (1998), the increased computing power in modern control systems permits higher level algorithms to be incorporated into the controllers. It is now possible to execute neural network algorithms as multi-input, multi-output functions in a DCS controller with full redundancy. An example of this is the Southern California Edison Mohave station neural network based sootblower system.

Neural networks will soon be embedded in the Human Machine Interface (HMI) level of control systems for improvements in alarm management. Root cause alarm development will be accelerated by neural network and hybrid expert system applications. Neural networks might be used to combine video information, such as thermographic imaging (infrared temperature detection) for remote diagnosis of devices that have limited instrumentation.

Appendix A
Mathematical Procedures

Within the body of this book there are references to a number of mathematical procedures which are reviewed in this appendix: multiple linear regression analysis, and several numerical analysis techniques, chief among them the Newton-Raphson method.

A.1 Multiple Linear Regression Analysis

Multiple linear regression analysis was originally conceived as a means for statistically analyzing experimental data. Draper and Smith (1968) provide a thorough outline of the fundamental theory, and reference should be made to their work. Consider a situation in which a result R can be measured corresponding to the measured values of a set of n independent variables $X_1, X_2, \ldots, X_n$. When m complete sets of (X,R) data are presented to a multiple linear regression analysis program, it will calculate the values of coefficients $a_0, a_1, \ldots, a_n$ in the following expression:

$$R - a_0(1.0) + a_1 X_1 \mid a_2 X_2 + \cdots + a_n X_n \qquad (A.1)$$

these values of a providing the best (minimum least-squares) fit of this expression to all the data sets.

In matrix notation, if the values of R in the m data sets are arranged as a vector $\mathbf{Y}$ while the values of X are arranged as an (X_{n+1}, m) matrix $\mathbf{Z}$ in which the first column contains the value

$X_0 = 1.0$ for all data sets, and if vector $\mathbf{A}$ is to contain the elements a_0, a_1,..., a_n, then:

$$A = (Z'Z)^{-1}Z'Y \qquad (A.2)$$

In matrix $\mathbf{Z}$, the number of rows should be at least equal to the number of variables plus 1, or $m \geq (n + 1)$; $\mathbf{Y}$ will contain m elements; and the vector $\mathbf{A}$ will contain $n + 1$ elements. Draper and Smith also indicate several criteria for judging the quality of the fit of the calculated array $\mathbf{A}$, but these are not of concern when regression analysis is being used within a system for monitoring condenser performance. The principal software modules required to resolve the value of vector $\mathbf{A}$ include a module to transpose a matrix, a module to multiply two matrices together, and a module to calculate the inverse of a matrix.

The main use of regression analysis in eqipment performance modeling is to generate the coefficients in polynomial relationships which take the form:

$$R = a_0(1.0) + a_1X + a_2X^2 + a_3X^3 \qquad (A.3)$$

Here R is again the result, while X is the value of a *single* independent variable associated with each value of R. Generating the $\mathbf{Z}$ matrix in Equation (A.2) so as to reflect the polynomial form of Equation (A.3) requires making the substitutions $X_1 = X$, $X_2 = X^2$, and $X_3 = X^3$. Equation (A.2) can now be used to solve for vector $\mathbf{A}$, which will contain a new set of values a_0, a_1,..., a_n that best satisfy the polynomial of Equation (A.3).

A.2 Numerical Analysis

With equations of the form $Y = f(X)$, the value of Y can be calculated directly from the selected value of X. However, if it is desired to calculate the value of X corresponding to the value of interest $Y = Y^*$, it will be necessary to use some type of numerical analysis method in order to resolve the value of X when $Y = Y^*$, unless the equation can be directly converted to the form

$$X = f(Y)$$

The literature contains several simple methods for solving $Y^* = f(X)$ described in the following subsections.

A.2.1 Fibonacci Search

A Fibonacci search can be useful if it is known with certainty that the solution value of X^* lies within a range L of Y whose upper and lower boundaries (Y_{max}, Y_{min}), together with the associated values of X_{max} and X_{min}, have been determined. Wilde (1964) shows the economy of a *search by golden section,* in which a first estimate of the solution is obtained by dividing the range L by the golden mean $\tau = 1.618033989$ and evaluating the value of X corresponding to the value of:

$$\text{Either } \left(X^* = X_{max} - \frac{L}{\tau} \right) \text{ or } \left(X^* = X_{min} + \frac{L}{s} \right) \quad \text{(A.4)}$$

The error corresponding to the value of X^* chosen for this first estimate is $(Y - Y^*)$, and so, depending on its sign, the upper and lower limits of either the longer or shorter section of L become the new values of X_{max} and X_{min}. A new value of X^* is calculated according to Equation (A.4) using the reduced length of L, and the new error $(Y - Y^*)$ is then calculated; the search process continues until the last calculated error value lies within the desired tolerance.

A.2.2 Regula Falsi Search

Kreyszig (1973) outlines the regula falsi method, which, again, requires a knowledge of the range of X within which the solution value ($Y = Y^*$) will be found. By calculating the values Y_{max} and Y_{min} corresponding to the values of X_{max} and X_{min}, a first approximation of X^* can be calculated by using triangulation:

$$X^* = \frac{X_{min}Y_{max} - X_{max}Y_{min}}{Y_{max} - Y_{max}} \tag{A.5}$$

A new value of Y^* can be calculated from X^* together with the error ($Y - Y^*$), and so, depending on the sign of the error, the newly calculated values of X^* and Y^* will be substituted for either the present upper or the present lower limit values of X and Y-i.e., for (X_{max}, Y_{max}) or for (X_{min}, Y_{min}). The updated range limits of X and Y will now be substituted in Equation (A.5) and new values of X^* and Y^* calculated, together with a new value of the error ($Y - Y^*$). The process continues until the error falls within the desired tolerance.

A.2.3 Newton-Raphson Numerical Search Technique

Kreyszig [1973, p. 764] cites a numerical analysis technique developed by Joseph Raphson [1697] that has been used in a variety of ways within programs designed to monitor condenser performance. Figure A.1 demonstrates the principles involved. In this example, the value of X is sought for the equation $Y = X^3 - 2X - 5 = 0$. The desired tolerance is an error less than 10^{-8}. X is chosen for first estimate to have a value of 1.5 (point A), and it is seen that the error at point A in Figure A.1 is -4.625 while the slope is 4.75. To obtain an improved estimate of the solution, the value of X should be changed by an amount equal to the quotient $-$error/slope. In this case, the change in X (DELX1, in the Figure) would be

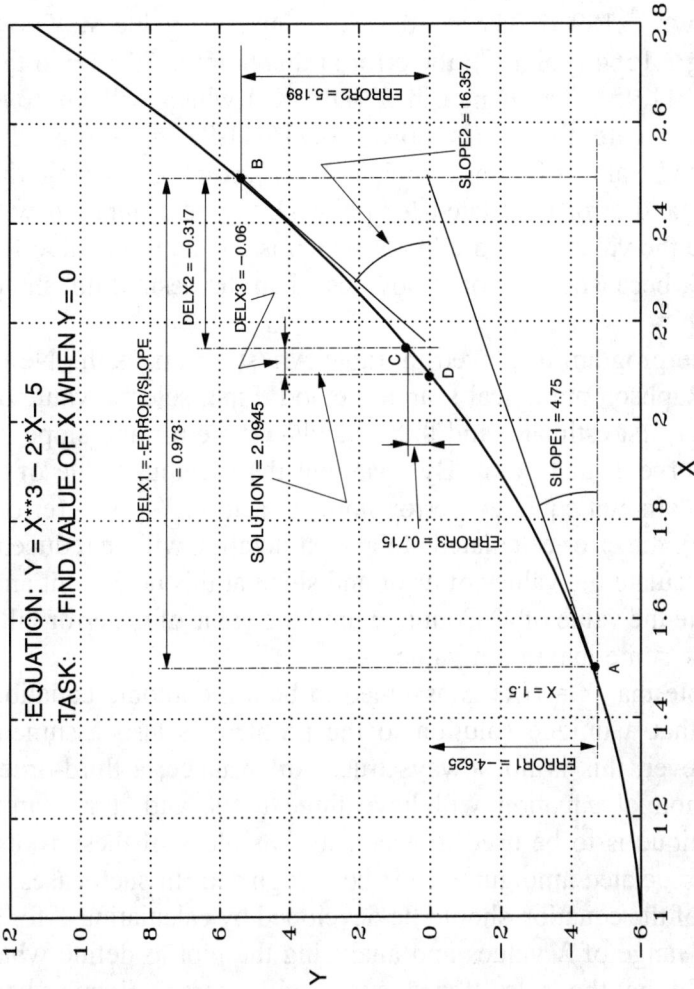

EQUATION: Y = X**3 − 2*X − 5

TASK: FIND VALUE OF X WHEN Y = 0

Figure A.1   Demonstration of Newton-Raphson method

0.973, so that the value of X which will provide an improved estimate of Y is now 2.473 (point B), the error at this point being 5.189 and the slope 16.357.

The change in X (DELX2) which will improve the estimate is now $-5.189/16.357 = -0.317$, the updated value of X now being 2.1564 (point C); the error at this point is 0.715 and the slope 11.95. The change in X (DELX3) which will improve the estimate further is now $-0.715/11.95 = -0.06$, the updated value of X now being 2.097 (point D). Using the slope and error calculated for point D, the next iteration will move the value of X to 2.0945, which is the solution value for $Y = 0$, because the error is now less than the desired threshold of 10^{-8}.

The program displayed in Table A.1 summarizes the Newton-Raphson numerical search method. First, select a value of X as a first estimate, and then calculate the error and slope of the curve at that point. By changing the previous value of X by an amount equal to $-$error/slope, or setting $X = X - $ (error/slope), the error calculated at the next iteration will be reduced. Recalculate the values of error and slope at this new point and update the value of X. Continue until the value of the error falls below some maximum value.

Note that $Y = f(X)$ is assumed to be a monotonic equation and that a unique solution to the problem is also assumed. However, this is not always true. For instance, a third-order polynomial equation will have three roots, and if a search technique is to be used to establish the values of these roots, the associated ambiguity must be recognized. In such cases, a plot of the equation should be developed by calculating Y for a wide range of X values and analyzing the plot to define what should be the rules for a successful search. Even when monotonicity is present, the search procedure can often be speeded up by first calculating X over a wide range of Y, noting when the error $(Y-Y^*)$ changes sign, and so establishing a reduced range of X within which to conduct the final search.

Table A.1   Program to verify the Newton-Raphson principle

```
INTEGER*2    I
REAL*8   X, ERRX, DFDX, DELX
FUNDAMENTAL EQUATION:   Y = X**3 - 2.0 * X - 5.0
TASK: FIND THE VALUE OF X WHEN Y = 0
ASSIGN INITIAL VALUE TO 'X' AND CALCULATE
ASSOCIATED ERROR
X = 1.5
  ERRX = X**3 - 2.0*X - 5.0
    WRITE(*,101)
      WRITE(*,100) X,ERRX
        WRITE(*,102)
          WRITE(*,103)
        I = 0
LOOP TO IMPLEMENT NEWTON-RAPHSON PRINCIPLE
      DO WHILE(ABS(ERRX) .GE. 1.0E-08)
        I = I + 1
          DFDX = 3.0 * X**2 - 2.0
            DELX = - 1.0 * ERRX / DFDX
              X = X + DELX
                ERRX = X**3 - 2.0*X - 5.0
                  WRITE(*,104) I,X,ERRX,DFDX,DELX
        END DO
WRITE(*,101)
SOLUTION
WRITE(*,200) X
PRINTOUT FORMATS
100 FORMAT(1H ,' Initial value of X = ',F15.9,' Initial Error =
', + F15.9 // )
101 FORMAT(20X / )
102 FORMAT(1H ,' Iteration Partial')
103 FORMAT(1H ,' No. X Error Differential + Delta X' / )
104 FORMAT(1H ,I5,4F15.9)
200 FORMAT(1H,'Solution to equation : X**3 - 2.0*X - 5.0 = 0.0 :
    + X = ', F12.9 // )
END
```

A.3 The Newton-Raphson Technique Used in Equipment Performance Models

A condenser performance model is used as an example of the application of the Newton-Raphson technique for solving sets of non-linear simultaneous equations. If there are n variables, there must also be n equations, some of which define the boundary conditions on which the model must converge. For instance, to model the behavior of a clean single-compartment condenser in a fossil fuel plant requires a matrix having 5 equations and 5 variables, the boundary conditions consisting of generated power, cooling-water inlet temperature, and flow (see Table A.2). At model convergence, all equations take the form:

$$f(X_1) + f(X_2) + \cdots + f(X_n) = 0 \qquad (A.6)$$

Once the set of equations has been defined, the n variables in the equation set are assigned initial values, which might be

Table A.2   Newton-Raphson data structures for single-compartment condenser

Change vector G					
$\Delta TFLOW$	ΔTIN	$\Delta EXH1$	$\Delta STMP1$	$\Delta TOUT1$	
1					$-\mathrm{err}(1)$
	1				$-\mathrm{err}(2)$
		1	$\partial f(3)/\partial STMP1$		$-\mathrm{err}(3)$
$\partial f(4)/\partial TFLOW$	$\partial f(4)/\partial TIN$	$\partial f(4)/\partial EXH1$		$\partial f(4)/\partial TOUT1$	$-\mathrm{err}(4)$
$\partial f(5)/\partial TFLOW$	$\partial f(5)/\partial TIN$		$\partial f(5)/\partial STMP1$	$\partial f(5)/\partial TOUT1$	$-\mathrm{err}(5)$
Matrix of partial differentials					Error vector E

the present set of operating conditions. Using this initial set of variable values, the sum of $f(X_i)$, $i = 1,\ldots,n$, can be evaluated for each equation, and, since it is unlikely that convergence has already occurred, each of these sums will have a nonzero value. These sums can be stored in a vector $\mathbf{E}$ as the *negatives* of their values.

An $n \times n$ matrix $\mathbf{F}$ can now be constructed from each equation, the matrix elements consisting of the partial differentials of each variable in each equation. These partial differentials can be determined in two ways:

1. By inspection of the equation. For example, if $f(X_{i,j}) = a_1 X_{i,1}$, then the partial differential of element $X_{i,j}$ will be a_1, which will be stored in element $F_{i,j}$.
2. Alternatively, the partial differential of $X_{i,j}$ with respect to equation j can be calculated by perturbing $X_{i,j}$ by a small amount, calculating the effect of that perturbation on the sum for equation j, and then dividing the change in the sum by the perturbation.

Assume a vector $\mathbf{G}$ which is to contain the *changes* to be made in the value of each variable in order for the solution to approach closer to convergence, convergence being defined as having occurred when all values contained in $\mathbf{G}$ fall below the threshold value assigned to each variable. Considering the tableau contained in Table A.2, the basic relationship between the matrix and the vectors is $\mathbf{GF} = \mathbf{E}$, from which the values for array $\mathbf{G}$ can be calculated using:

$$\mathbf{G} = \mathbf{EF}^{-1} \tag{A.7}$$

After updating the set of variables by applying to each the corresponding 'change' value contained in the elements of the $\mathbf{G}$ vector, new values are calculated for each equation and stored in array $\mathbf{E}$, while all the partial differentials are also recalculated, using the updated values of each variable. The procedure defined in Equation (A.7) is performed and a new set of change values calculated. If the absolute values of all changes fall below the assigned thresholds, then convergence is declared and the current set of X values becomes the solution. Otherwise, if the change to be made in any variable should fall outside its assigned tolerance, the iterative procedure will continue until convergence takes place. In practice, convergence for a condenser matrix will occur in fewer than 10 iterations. Should a larger number of iterations be required, it is possible that some tolerances will need to be adjusted or the set of equations critically examined.

There are strong similarities between this method of solving a set of nonlinear simultaneous equations and the Newton-Raphson search technique outlined in Section A.2.3. In both cases, in order to cause the model to move closer to convergence, a better variable, or set of variables, can be obtained by changing the present values of X_i, $i = 1,\ldots, n$, by amounts equivalent to the negative of the error divided by the slope or, in the case of condenser models, the matrix equivalent of slope considered as the set of partial differentials.

A.4 Linear Programs

Linear programming has long been a useful operations research tool for the analysis of networks and their optimization. The widespread availability of computers and the development of easy to use software now allow this technique to be used successfully by industrial plants to solve a large number of practical problems, and without needing a deep understanding of the mathematical sophistication included in the design of the algorithm. Much of the original literature (Dantzig (1963) and Hadley (1962)) was written when industrial computers were in their infancy but their possibilities for reducing this theory to a widespread industrial practice were obvious. The frequent application that emerged in the late seventies was for optimizing the energy distribution network within cogeneration plants and, while the cost of energy remained high, numerous real-time computer control systems were installed in steel, paper and chemical processing plants and successfully reduced the quantity of energy consumed by those plants as well as its cost.

The model used in a linear program is a network of energy resource variables inter-related to their points of consumption by a set of linear equations that are in the form of both equalities and inequalities. The network is also provided with a cost function that defines the unit cost of each purchased resource and uses this to determine that distribution of resource assignments that will minimize the purchased energy costs in order to satisfy the plant energy demands. Thus the needs of the production departments are paramount and the task of the energy department is, first of all, to satisfy these needs at least cost. However, there may be times when these needs exceed the present capacity of the available energy resources, and the task then is to advise the production departments of the constraints that are being encountered. There is thus not only an on-line real-time optimization problem to be solved but also an advisory function to be performed should plant energy needs encounter one or more system operating constraints.

Dantzig (1963) and Hadley (1962) both contain full details of the mathematical features contained in linear programming algorithms and should be referred to. The purpose of this section is to draw attention to those features that affect the application of these algorithms to solve practical problems and that the user should bear in mind. Consider a simple network problem defined in the form of three inequality equations and a cost function:

$$2x_1 + 3x_2 \leq 6 \tag{A.8}$$

$$x_1 + 7x_2 \geq 4 \tag{A.9}$$

$$x_1 + x_2 = 3 \tag{A.10}$$

and a cost function:

$$4x_1 + 3x_2 = z \tag{A.11}$$

To transform the inequalities into linear equations, artificial variables are added. Those associated with upper constraints (Equation (1)) are assigned a value of +1 while those associated with lower constraints (Equation 2) are assigned a value of −1 to produce the following starting tableau Table A.4.1:

Note that it is assumed that all variables in the solution will be either zero or have non-negative values. It should also be noted that the basic algorithm is usually structured to solve a maximization problem. To solve a minimization problem the cost function is multiplied internally by −1. The algorithm is usually structured in the Simplex or Revised Simplex form.

In addition to searching for an optimal solution to the network problem presented to it, the linear programming algorithm first checks whether there is in fact a feasible solution to the problem. If there is no feasible solution it can mean that redundant equations have been defined and one or more must be removed. Alternatively, that there are errors in

Table A.4.1   Linear programming tableau

| Equation | Energy Variables | | Artificial Variables | | |
	x1	x2	x3	x4	Constant
1	2	3	1	0	6
2	1	7	0	−1	4
3	1	1	0	0	3
Cost Function	4	3	0	0	

certain equations or they are inconsistent with one another. One form of non-feasible solution occurs if the demand for energy exceeds the upper constraints on one or more resources. In this case, the problem may need to be restructured so that the plant energy demands are defined as lower constraints rather than equalities and the program run in a "what-if" mode to explore where the true resource constraint(s) lie.

Once the algorithm has determined that there is a feasible solution it proceeds to iterate through the matrix until it achieves convergence and then displays the results. The significant values are those of the energy variables. If some of the artificial variables have non-negative values at convergence these may be ignored.

Occasionally the algorithm reports that the solution is unbounded. If this should be the case additional constraints have to be added to provide the necessary boundary.

All of this illustrates the importance of exercising the network model off-line to obtain a good understanding of the results that will be obtained from a wide variety of combinations of plant operating conditions, as they will be applied to the network. Clearly, the control system associated with real-time on-line control applications must be protected from the effects of encountering any non-feasible solutions, which should be considered an alarm condition. If encountered, the control system set points should be switched to manual and the problem studied using the off-line version of the linear program.

A.4.1 Discontinuous Linear Relationships

It is occasionally necessary to incorporate discontinuous linear relationships within a linear programming context and the following outlines the matrix structure that can be used to reflect such relationships (See Dantzig (1963) pp. 484–486). The construct allows the automatic inclusion of all those segments that apply to the present value of the independent variable and relies on the fact that, at the convergence of a linear program (LP), all variables must have either a positive value or be zero.

Consider a discontinuous relationship between Nox and generated power that is shown in Figure A.2. It consists of three linear segments:

$$NOX1 = 20 - 0.1667^*MW \qquad 0 \leq MW \leq 100 \qquad (A.12)$$

$$NOX2 = 0.2^*(MW - 45) \qquad 45 \leq MW \leq 75 \qquad (A.13)$$

$$NOX3 = 0.1^*(MW - 75) \qquad 75 \leq MW \leq 100 \qquad (A.14)$$

$$NOX \ = NOX1 + NOX2 + NOX3 \qquad (A.15)$$

Equation (A.12) is an equality and is continuous over the whole MW range. Equation (A.13) represents the increase in Nox as a function of MW over the load range of 45 to 75 MW; while Equation (A.14) represents the *additional* increase in Nox over the load range of 75 to 100 MW. Prior to being inserted into a LP matrix, the equations must be transformed as follows:

$$-0.1667^*MW - NOX1 = -20 \qquad (A.16)$$

$$0.2^*MW - NOX2 \leq 9 \qquad (A.17)$$

$$0.1^*MW - NOX2 \leq 7.5 \qquad (A.18)$$

The corresponding LP matrix structure is shown in Table A.4.2.

Note that only the Nox variable is assigned a cost value and that the LP matrix is to BE solved by searching for the minimum value of Nox.

To test that the above structure reflects the plot of Figure A.2 the above linear programming matrix was exercised for

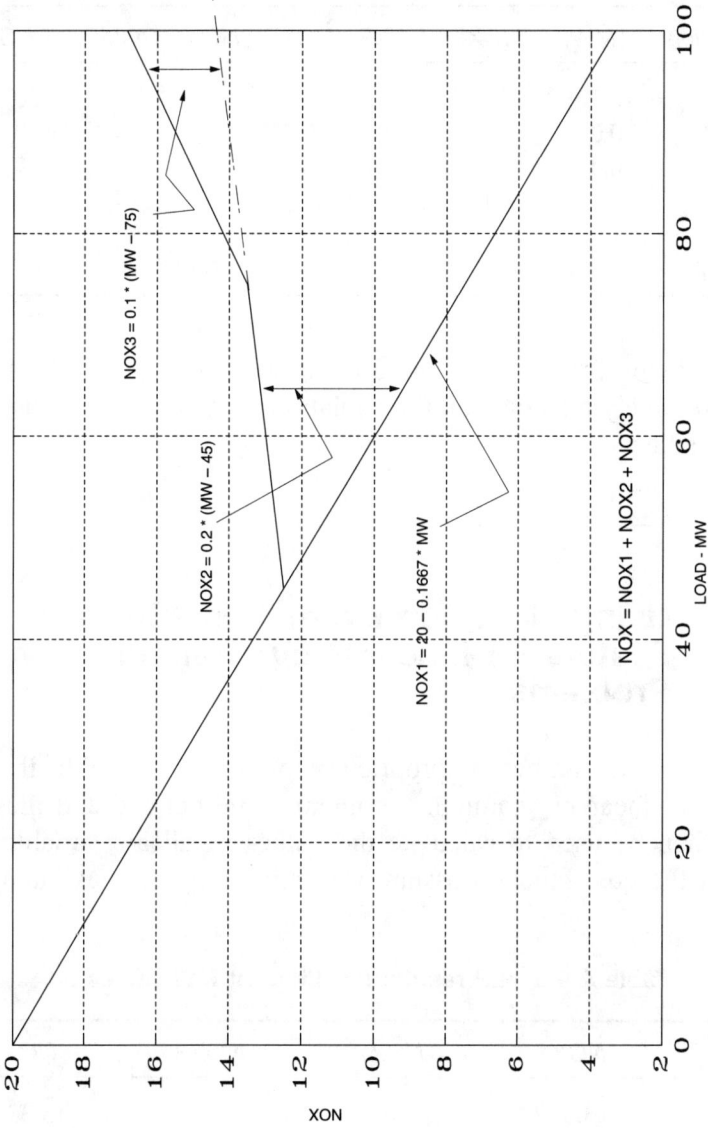

NOX3 = 0.1 * (MW −75)

NOX2 = 0.2 * (MW − 45)

NOX1 = 20 − 0.1667 * MW

NOX = NOX1 + NOX2 + NOX3

LOAD - MW

XON

Figure A.2   Discontinuous relationship

Table A.4.2   LP matrix to reflect a discontinuous linear relationship

Equation	MW	NOX1	NOX2	NOX3	NOX	COMP	RHS
5a	−0.1667	−1.0	−1			=	−20
5b	0.2					≤	9
5c	0.1		−1			≤	7.5
5d		1	1	1	−1	=	0
MW	1					=	80
COST					0.5		

MW having the values of 40, 50 and 80 MW respectively, so progressively including all three relationships. Table A.4.3 lists the results:

A.5   Using MicroSoft Excel™ as "Solver" Feature for Linear Programming Problems

There are a number of proprietary programs available that allow a linear programming problem to be defined and they can then be used to calculate the values of all the variables when the cost function assumes its minimum (or maximum)

Table A.4.3   LP results for different MW values

MW	NOX1	NOX2	NOX3	NOX
40	13.332	0	0	13.332
50	11.665	1	0	12.665
80	6.664	7	.5	14.164

value. This Appendix provides some notes for solving linear programming problems using the Solver feature in Microsoft EXCEL™. To illustrate the method, reference is made to the flow sheet for a small power house contained in Figure A.2.

In this flow sheet there are nine energy resources, namely:

1. Steam turbogenerator: throttle flow   lb/h
2. Steam turbogenerator: extraction flow   lb/h
3. Steam turbogenerator: condenser flow   lb/h
4. Steam turbogenerator: power   kw
5. HP/IP Press red. Valve flow   lb/h
6. IP/Process steam Press red. Valve flow   lb/h
7. HP boiler flow   lb/h
8. IP boiler flow   lb/h
9. Tie-line power   kw

It is customary to define a linear programming problem in the form of a matrix, as shown in Table A5-I. The columns X.1 thru X.9 represent the energy resources. To the right of these columns is a column indicating the *nature* of the equality or inequality ($<=$, $=$, $>=$); and to the right of that is the *value* of the equality or inequality.

The first line contains the cost function that contains the unit prices of only the *purchased* energy resources. The internal prices of the other resources are irrelevant.

The rows Y.1 thru Y.19 contain the relationships between the energy resources, expressed as a set of equality or inequality equations. In this case, these are as follows:

A.5.1 Upper Constraints

Y.1 Steam turbogenerator: throttle flow   lb/h
Y.2 Steam turbogenerator: extraction flow   lb/h
Y.3 Steam turbogenerator: condenser flow   lb/h
Y.4 Steam turbogenerator: power   kw

Table A5-I Linear programming matrix

| CHAP04A | OBJECTIVE: MIN | | VARIABLES: 9 | | DATE 08-14-2001 | | | |
| BASIS: NONE | CONSTRAINTS: 17 | | SLACKS: 11 | | TIME 15:07:08 | | | |

	X.1	X.2	X.3	X.4	X.5	X.6	X.7	X.8	RHS
MIN COST							.0050	.0045	.000000
Y.1	1.000								<=180000
Y.2		1.000							<=120000
Y.3			1.000						<=140000
Y.4				1.000					<=9500.00
Y.5							1.000		<=200000
Y.6								1.000	<=100000
Y.7									<=3800.00
Y.8	1.000	-1.000	-1.000						=.000000
Y.9	-1.000								=.000000
Y.10					-1.000	-1.000	1.000		=40000.0
Y.11		1.000			1.130	1.000			=30000.0
Y.12				1.000					=7000.00
Y.13			-.5000	-8.700					=7200.00
Y.14	1.000		1.000						>=10000.0
Y.15									>=2000.00
Y.16							1.000		>=20000.0
Y.17								1.000	>=10000.0

CHAP04A BASIS: NONE	OBJECTIVE: MIN CONSTRAINTS: 17	VARIABLES: 9 SLACKS: 11	DATE 08–14–2001 TIME 15:07:28

MIN	X.9	RHS
COST	.0400	.000000
Y.1		<=180000
Y.2		<=120000
Y.3		<=140000
Y.4		<=9500.00
Y.5		<=200000
Y.6		<=100000
Y.7	1.000	<=3800.00
Y.8		=.000000
Y.9		=.000000
Y.10		=40000.0
Y.11		=30000.0
Y.12	1.000	=7000.00
Y.13		=7200.00
Y.14		>=10000.0
Y.15		>=2000.00
Y.16		>=20000.0
Y.17		>=10000.0

Y.5 HP boiler load   lb/h
Y.6 IP boiler load   lb/h
Y.7 Tie-line power   kw

A.5.2 Equalities

Y.8   Turbine steam balance: $X1 - X2 - X3 = 0.0$
Y.9   HP steam balance: $-X1 - X5 + X7 = 0.0$
Y.10 IP Steam to process: $1.13*X5 - X6 + X8 = 40000$
Y.11 LP steam to process: $X2 + X6 = 30000$
Y.12 Power balance: $X4 + X9 = 7000$
Y.13 Steam turbogenerator characteristic curve: $X1 - 0.5*X2 - 8.7*X4 = 7200$

A.5.3 Lower Constraints

Y.14 Steam turbogenerator Condenser flow   lb/h
Y.15 Steam turbogenerator: power   kw
Y.16 HP boiler load   lb/h
Y.17 IP boiler load   lb/h

When this matrix is executed using a proprietary linear programming algorithm, the minimum cost is calculated to be $579.94/hour, the values of the various energy resources at minimum cost being shown in Table A5-II.

A.5.4 Creating the Excel™ Spreadsheet

To use the MicroSoft EXCEL™ Solver feature to solve a linear programming problem, an EXCEL™ spreadsheet must first be created, as shown in Table A5-III. In this table, the names of the nine energy resources are inserted in column A on lines 5 thru 13. A zero is then inserted in cells C5 thru C13. These cells will eventually receive the minimum cost solution values after the optimization problem has been solved.

Table A5-II   Results at minimum cost

CHAP04A	SOLUTION IS MINIMUM PRIMAL PROBLEM SOLUTION	COST	579.9433628		DATE 08–14–2001 TIME 15:08:13
VARIABLE	STATUS	VALUE	COST/UNIT	VALUE/UNIT	NET COST
X.1	BASIS	50040.000	.00000000	.00000000	.00000000
X.2	BASIS	30000.000	.00000000	.00000000	.00000000
X.3	BASIS	20040.000	.00000000	.00000000	.00000000
X.4	BASIS	3200.0000	.00000000	.00000000	.00000000
X.5	BASIS	26548.673	.00000000	.00000000	.00000000
X.6	NONBASIS	.00000000	.00000000	–.0192478	.0192478
X.7	BASIS	76588.673	.005000000	.005000000	.00000000
X.8	BASIS	10000.000	.00450000	.00450000	.00000000
X.9	BASIS	3800.0000	.040000000	.040000000	.00000000

Table A5-III   Original spread sheet

A	B	C	D	E	F	G	Resource ID	Line ID
CHAP04AA L.P. PROBLEM								
Throttle flow		0	180000	0			1	5
Extraction flow		0	120000	0			2	6
Condenser flow		0	140000	10000			3	7
Power		0	9500	2000			4	8
HP red. Valve		0	50000	0			5	9
LP red. Valve		0	50000	0			6	10
HP boiler		0	200000	20000	0.005		7	11
LP boiler		0	100000	10000	0.0045		8	12
Tie-line		0	3800	0	0.04		9	13
Objective Function			<=	>=	cost			16
C5–C6–C7=0						0		18
–C5–C9+C11=0						0		19
1.13*C9–C10+C 12=40000						40000		20
C6+C10=30000						30000		21
C8+C13=7000						7000		22
C5–0.5*C6–8.7*C8=7200						7200		23

Cells D5 thru D13 should contain the values of the upper constraints for all variables while cells E5 thru E13 should contain the values of the lower constraints for all variables. Note that, in the absence of a positive value the lower constraint should be zero. This is consistent with the LP axiom that the values of all variables in the solution must be >= zero.

The values stored in the appropriate cells in column F are the prices of each of the purchased resources only.

Cell G16 is assigned to the cost function. The equation for calculating the cost function should be loaded into this cell. In this case the cost function is: C11*F11+C12*F12+C13*F13.

Cell G18 thru G23 contain the equality equations listed above as equations Y.8 thru Y.13. The left hand sides of these equations should be loaded into these cells.

A.5.5 Completing Problem Initializing Using "Solver"

Click on "Tools" on the toolbar and then click on "Solver". The Solver Parameter Wizard will appear. The following data has to be inserted in the format given below:

A. Set target cell: $G$16
B. Click on "Min" to set minimization problem
C. "By changing cells": $C$5:$C$13 The colon separating the two cells indicates a range of cells
D. "Subject to the constraints": Click on "Add" and enter all the constraints. These include upper constraints, lower constraints and all equalities.
   Typical upper constraint: $C$5<=$D$5
   Typical lower constraint: $C$5>=$E$5
   Typical equality constraint: $G$20=40000

E. To solve the problem with the present set of data, click on "Solve" within the "Solver Parameter Wizard". The results will be displayed as:
Variables: stored in cells C5 thru C13,
Cost per hour: cell G16
Constraints: cells G18 thru G23.

A.5.6 Testing the Program

Given the energy demands by the plant, in this case in terms of the IP steam consumption (G20), the LP steam consumption (G21) and total power consumption (G22), the optimal solution provided by the linear programming algorithm provides the cost per hour for the problem (G16) and the values of all the energy resource variables when solution has the minimum cost (C5 thru C13). Changing any or all of these three consumption targets will change the minimum cost as well as the values of some or all of the resources.

References

Allen, Thomas and Michael Polito (1992), "Energy Management Software for a Combined Cycle Cogeneration Facility," International Turbomachinery, Vol. 33, No. 1, January/February 1992.

ASME Power Test Code PTC.4.1-1974 (1974), "Steam Generating Units," ASME, New York, NY.

ASME Power Test Code PTC.4-1998 (1998), "Fired Steam Generators," ASME, New York, NY.

ASME Steam Tables (1967), 6th Edition (1993), publ. American Society of Mechanical Engineers, New York, NY.

Åstrom, K.J., U. Borisson, L. Ljung and B. Wittenmark (1977), "Theory and Application of Self Tuning Regulators," Automatica, Vol. 1, No. 3, pp. 457–476.

Baldwin, Neil (1995), Edison: Inventing the Century, Hyperion, New York.

Bowman, R. A., Mueller, A. C. and Nagle, W. M. (1940), "Mean Temperature Difference in Design," ASME Transactions, Vol. 62, pp. 283–294.

Box, G. E. P. and J. S. Hunter, (1959), "Condenser Calculations for Evolutionary Operation Programs," Technometrics, Vol. 1, No. 1 February 1959, pp. 77–95.

Box, G. E. P. (1957), "Evolutionary Operation: A Method for Increasing Industrial Productivity," Applied Statistics, Vol. 6, 1957, pp. 3–22.

Brownlee, William D. and Lemanowicz, James M. (2001), "Open Standards Energize Powerplant Controls," POWER, July/August 2001, pp. 52–56.

Buna, T. (1956), "Combustion Calculations for Multiple Fuels," Transactions of the ASME, August 1956, pp. 1237–1249.

Carpenter, B. H. and Sweeney, H. C. (1965), "Process Improvement with 'Simplex' Self-Directing Evolutionary Operation," Chemical Engineering, July 5, 1965, pp. 117–126.

"Cooling Tower Manual" (1976), Chapter 3, Cooling Tower Performance Variables. Cooling Tower Institute, June 1976.

"Cooling Tower Manual" (1977), Chapter 2, Basic Concepts of Cooling Tower Operations. Cooling Tower Institute, January 1977.

"Cooling Tower Manual" (1981), Chapter 1, Cooling Tower Operations. Cooling Tower Institute, January 1981.

CTI Code Tower (1990), Acceptance Test Code for Water-Cooling Towers, Cooling Tower Institute, Houston, TX.

Dantzig, A. B., (1963), "Linear Programming and Extensions," Princeton University Press, 1963, p. 485.

Dantzig, George B. (1963), Linear Programming and Extensions, publ. Princeton University Press, Princeton, NJ.

Draper, N. R. and H. Smith (1968), *Applied Regression Analysis*, John Wiley & Sons, New York.

Eggert, C. W. (1967), "Evolutionary Operation (E.V.O.P.) – A Promising Technique in Minreal Dressing," The Canadian Mining and Metallurgical Bulletin, October 1967, pp. 1169–1172.

Elgerd, O. I. (1971), *Electric Energy Systems Theory: An Introduction,* McGraw Hill Company, NY, 1971.

Elleson, Jim (2000), "HVAC&R Newsletter", The EPRI HVAC&R Center Quarterly Newsletter, March 2000.

Goff, John A. and S. Gratch (1945), "Thermodynamic Properties of Moist Air," ASHRA Transactions, Vol. 51, pp. 125–164.

Hadley, G., (1962), *"Linear Programming,* publ. Addison-Wesley Publishing Company Inc." Reading, MA.

Hanway, James E. and Richard E. Putman (1992), U.S. Patent No. 5,081,591, "Optimizing Reactive Power Distribution in an Industrial Power Network."

HEI (1980), *Standard for Power Plant Heat Exchangers,* 1st ed., Heat Exchange Institute, Cleveland.

Hewitt, Shires and Bott (1994), "Process Heat Transfer," CRC Press, 1994, pp. 578–579.

Hirst, A. W. (1942), *Applied Electricity,* Blackie and Sons London.

Hooke, R. and T. A. Jeeves (1961), "'Direct Search' Solution of Numerical and Statistical Problems," J. Assoc. Comp. Mach., vol. 8, No. 2, April 1961, pp. 212–229.

Horlock, J. H. (1992), "Combined Power Plants," Pergamon Press, Oxford, UK.

Immonen, Pekka J. (2000), "Mathematical Models in Cogeneration Optimization", Instrument Society of America POWID Newsletter, December, 2000.

Kalman, R. E. (1964), "When is a Linear Control System Optimal?", Journal of Basic Engineering, Vol. 86, pp. 71–80.

Katebi, M. et al. (1995), "New Software Tools for Power Plant Modeling, Simulation and Optimization," Proc. Power-Gen Europe, Amsterdam, May 16–19th, pp. 131–164.

Kehlhofer, Rolf (1991), "Combined Cycle Gas and Steam Turbine Power Plants," Fairmont Press, Liburn, GA.

Kern, D. Q. (1958), *Process Heat Transfer,* 2nd Edition, McGraw Hill, New York.

Kreyszig, E. (1973), *"Advanced Engineering Mathematics",* John Wiley and Sons, New York.

Lewitt, E. H. (1953), "Thermodynamics Applied to Heat Engines," Sir Isaac Pitman and Sons, Ltd., London.

Lichtenstein, Joseph (1943), "Performance and Selection of Mechanical Draft Cooling Towers", Trans ASME, Vol. 65, pp. 779–787, April 1943.

Lovins, Amory B. and L. Hunter (2001), "Fool's Gold in Alaska," Foreign Affairs, July/August 2001, pp. 72–85.

Minkowycz, W. J. and Sparrrow E. M. (1966), "Condensation Heat Transfer in the Presence of Noncondensables, Interfacial Resistances, Superheating, Variable Properties, and Diffusion", Journal of Mass Heat Transfer, Vol. 9, pp. 1123–1144, Pergamon Press 1966.

Mollier, R. (1923), "Ein Neues Diagramm fur Dampfluftgemische", Z.V.D.I., Vol. 67, page 869, 1923.

Nagle, W. M. (1933), "Mean Temperature Differences in Multipass Heat Exchangers," Industrial Engineering Chemistry, Vol. 26, pp. 604–608.

O'Keefe, William (1986), "How to Put Together Systems for Today's Desuperheater Needs," POWER, January 1986, pp. 13–20.

Ordys, A. W. et al. (1994), "Modeling and Simulation of Power Generation Plants," Springer-Verlag, London.

Peressini, A. L. et al. (1988), "The Mathematics of Non-linear Programming," Springer-Verlag, New York.

Preheim, Joel and Abruzere, Gene (1998), "Automation Links Remote Sites," Power Engineering, September 1998, pp. 35–39.

Putman Richard E. (1986), U.S. Patent No. 4,628,462, "Multiplane Optimization Method and Apparatus for Cogneration of Steam and Power."

Putman, R. E. J. and K. A. Panizza (1984), "The Increased Digitizing of Process Control Systems," Proc. I.S.A. Productivity Symposium, Eugene, OR, September 1984.

Putman, Richard E. (1975), U.S. Patent No. 3,872,286, "Control System and Method for Limiting Power Demand of an Industrial Plant."

Putman, Richard E. (1978), "The Optimization of Non-linear Power Plant Systems," Proc. IASTED Energy Systems Conference, Montreal, June 1978.

Putman, Richard E. (1989), U.S. Patent No. 4,805,113, "Economical Dispatching Arrangement for a Boiler System Having a Cogenerative Capability."

Putman, Richard E. et al. (1992), U.S. Patent No. 5,159,562, dated Oct. 27, 1992, "Optimization of a Plurality of Multiple-Fuel Fired Boilers Using Iterated Linear Programming."

Putman, Richard E. (2001), "Steam Surface Condensers: Basic Principles, Performance Monitoring sand Maintenance," publ. ASME Press, New York, NY.

Putman, Richard E., Frederick C. Huff and Jayanta K. Pal (1999), "Optimal Reactive Power Control for Industrial Power Networks, IEEE Transactions on Industry Applications," Vol. 35, No. 3, May/June 1999, pp. 506–514.

Putman, Richard E., Ronald J. Paterni and Edward R. Kuzniarski (1996), "The Steady-State Modelling and Constrained Optimization of Combined Cycle Plants," Proc. IJPGC 1996, PWRE-Vol 30, pp. 367–381.

Raphson, J. (1697), *Analysis Aequationum Universalis*, 2nd Edition, Royal Philosophical Society, London.

Smith, C. E. and Putman, R. E. (1983), "Paper Mill Energy Management Using Distributed Control," Proceedings 1983 ISA Conference, pp. 1105–1116.

Smith, C. E. and R. E. J. Putman (1988), "Paper Mill Energy Management Using Distributed Process Control," Proc. I.S.A. International Conference, Houston, TX, October 10–13, 1988.

Spendley, W., G. R. Hext and F. R. Himswworth (1962), "Sequential Application of Simplex Designs in Optimization and Ecolutionary Operation," Technometrics, Vol. 4, No. 4, November 1962, pp. 441–460.

Swirski, Konrad and Williams, Jeffery J., "Closed Loop NOx Control and Optimization using Neural Networks", Instrument Society of America POWID Conference, June 1998.

TEMA (1988), *Standards of the Tubular Exchanger Manufacturers Association*, 7th ed., Tarrytown, NY.

Underwood, A. G. V. (1934), "The Calculation of the Mean Temperature Difference in Multi-Pass Heat Exchangers", *Journal of the Institute of Petroleum Technology*, Vol. 25, pp. 145.

Weisbrod, Glen and Ford, Ellen (1996), Proceedings of the Electric Power Research Institute's Innovative Pricing Conference, San Diego, March, 1996.

Wexler, Arnold and Richard Hyland (1981), "Thermodynamic Properties of Dry Air, Moist Air and Water and SI Psychrometric Charts," Report on ASHRAE Research Projects 216-RP and 257-RP, publ. National Bureau of Standards, 1981.

Wilde, Douglass, J. (1964), "Optimum Seeking Methods," Prentice-Hall, Inc., Englewood Cliffs NJ, pp. 123–158.

Wood, B. and Betts, P. (1950), "A Temperature-Total Heat Diagram for Cooling Tower Calculations," The Engineer (U.K.), March 17, 1950, pp. 337–339 and March 24, 1950, pp. 349–351.

Yeager, Bob (1997), "Control Meets the Enterprise," Industrial Computing, October 1997, pp. 24–26.

Zink, John C., (1999), "Information and Control Merge in New Environment", Power Engineering, June 1999, pp. 19–22.

Index

A

Adiabatic expansion	40
Air property algorithms	75
Asynchronous Transfer Mode (ATM)	14

B

Backpressure	50
Base loads	241
Bayer process	225
Boiler combustion efficiency	34

C

Cogeneration	49
Combined cycle plants	163
Combined heat and power plants (CHP)	50
Combustion	
control systems	2
efficiency	31
efficiency vs. load	139
optimization	259
properties of major fuel elements	35
Computer-Aided Design (CAD)	8
Condenser	44
Condenser back pressure	57
Condenser/cooling tower subsystem	190
Condenser duty	61
Condenser model	191
Control Variables (CVs)	261

Cooling tower characteristic curve	72
Cooling tower model	191
Cooling towers	66
Copper Distributed Data Interface (CDDI)	14
Cost function	87
Critical loads	241

D

DC excitation generator	196
Demand charge	79
penalty	236
Demand control system strategy	243
Demand period	237
Demand-side management	235
Desuperheating	44
water/steam flow ratio	83
Discontinuous linear relationships	277
Distributed Control System (DCS)	11
Distribute Control System (DCS)	7, 23
Duct burners	168
Dynamic Data Exchange (DDE)	26

E

eDB historian	20, 28
Efficiency vs. load curves	33
Electrical load	240

Electrical tariffs   236
Equal incremental cost
    method   100
Equal incremental cost
    techniques   80
Equipment models   170
Ethernet (IEEE 802.3
    standard)   14
Evaporative cooling   68
Expansion line   57
Extraction/condensing
    steam turbogenerator   53
Extraction governors   57

F
Fast ethernet   14
Fiber Distributed Data
    Interface (FDDI)   14
Fibonacci search   267
Film heat transfer coefficient   48
Five heat losses   37
Fouling   65

G
Gas turbogenerator   170
Geographical Positioning
    System (GPS)   14

H
Heat exchangers   213, 225
    process   233
    with U-tubes   233
Heat rate vs. load curve   102
Heat recovery steam
    generator   143, 174
Heat transfer coefficient   61
Higher Heating Value
    of the Fuel (HHV)   32
Hydrodrilling heat
    exchanger tubes   231

Hydrodrilling and
    high pressure water   233
Hydroelectric
    turbogenerators   154

I
Impedance   148
Incremental cost   39
Industrial boilers   31
Industrial energy system   77
Information technology   79
Inhibited loads   241
Injection of steam   168
Isochronous   204
Isothermal expansion   43

K
Kalman filter   107

L
L. P. matrix   121
LaGrange multiplier   100
Linear programs   80, 274
Linear programming
    matrix   83
    technique   111
Log Mean Temperature
    Difference (LMTD)   218

M
Maintenance Manage
    Systems (MMS)   23
Manipulated Variables (MVs)   261
Maximum Continuous
    Rating (MCR)   32
Maximum excitation current   208
Maximum winding current   198
Mechanical cleaners   62
Midland cogeneration venture   186
Multiple linear regression   265

N

National energy policy	4
NetDDE	27
Network Dynamic Data Exchange (NetDDE)	20
New base case	95
Newton-Raphson numerical search technique	148, 268
Neural net	260
Neural networks	258, 260
Nox emissions	168
Nox steam	168
Numerical analysis	266

O

Off-line regression	107
OLE (Object Linking and Embedding for Process Control (OPC)	20, 24
On-line regression analysis	106
Open architecture control system	15, 20
Open Data Base Connectivity (ODBC)	20, 22
Optimal unit load vs. total plant load	105
Optimization of VARS	142
Optimizing the utilization of energy	78
Output-input efficiency	33
Overall unit heat rate	50

P

P-50 industrial process computer	10
Power factor	198
Programmable Logic Controllers (PLC)	8

R

Rankine cycle	215
Reactive capability curves	147
Reactive power	144
Real power	144
Real time electrical power pricing	255
Real-time pricing	256
markets for	257
Recovery boilers	80
Reference bus.	144
Regula falsi search	268
Relational Database Management Systems (RDBMS)	8

S

Sheddable loads	242
Shedding of electrical loads	235
Shell and tube heat exchangers	217
Shell and tube type	215
Simple turbogenerator	49
Simples Self-directing EVOP	80
Simplex principle	92
Simplex Self-directing Evolutionary Operation (SSDEVOP)	
experimental design	93
optimizing technique	90, 135
procedure	97
solution of non-linear (boiler problem)	138
Single pass condenser	59
Sliding window	3, 239
Speed of the machine	196
Standard Query Language (SQL)	19
Static Load Flow Analysis (SLFA)	148
Stator	196
Steam desuperheaters	43, 45

Steam pressure
  reducing valve   40
Steam surface condensers   59
Steam turbogenerators   49
Steepest ascent methods   80, 98
Structure Query
  Language (SQL)   22
Synchronous mode   252

T
Tap-changing transformers   144
Temperature effectiveness   220
Three-phase voltages   196
Tie-line frequency   196
Tie-line power factor   144
Time of Use (TOU)   255
Transformation ratio   148
Transmission Control
  Protocol/Internet
    Protocol TCP/IP   14
Tube bundle   215
Tube mean metal temperature   222
Turboalternator   198
Turbogenerator

excitation   193
gross unit heat rate   50
heat rate   159
throttle flow equation   96
net unit heat rate   50
Two-pass heat exchanger   219

U
Under-cooling of the
  condensate   192
Unit heat rate   50, 159

V
VARS   144
Volumetric properties of
  major fuel elements   36

W
WAVE server   21
Web Access View
  Enabler (WAVE)   20
Wide Area Network (WAN)   17
Willans line   51
Windings of the rotor   196